Johannes Maximilian Meckbach

Superconducting Multilayer Technology for Josephson Devices

Karlsruher Schriftenreihe zur Supraleitung   Band 013

HERAUSGEBER

Prof. Dr.-Ing. M. Noe
Prof. Dr. rer. nat. M. Siegel

Eine Übersicht über alle bisher in dieser Schriftenreihe
erschienene Bände finden Sie am Ende des Buchs.

Superconducting Multilayer Technology for Josephson Devices

by
Johannes Maximilian Meckbach

Dissertation, Karlsruher Institut für Technologie (KIT)
Fakultät für Elektrotechnik und Informationstechnik, 2013
Hauptreferent: Prof. Dr. Michael Siegel
Korreferent: Prof. Dr. Reinhold Kleiner

Impressum

 Scientific
Publishing

Karlsruher Institut für Technologie (KIT)
KIT Scientific Publishing
Straße am Forum 2
D-76131 Karlsruhe

KIT Scientific Publishing is a registered trademark of Karlsruhe
Institute of Technology. Reprint using the book cover is not allowed.

www.ksp.kit.edu

Print on Demand 2013

ISSN 1869-1765
ISBN 978-3-7315-0122-0

Superconducting Multilayer Technology for Josephson Devices

zur Erlangung des akademischen Grades eines

Doktor-Ingenieurs

von der Fakultät für
Elektrotechnik und Informationstechnik
des Karlsruher Instituts für Technologie (KIT)

genehmigte

Dissertation

von

Dipl.-Ing. Johannes Maximilian Meckbach

geb. in Heidelberg

Tag der mündlichen Prüfung: 24. September 2013

Hauptreferent: Prof. Dr. Michael Siegel
Korreferent: Prof. Dr. Reinhold Kleiner

Preface

The work presented within this dissertation has mainly been carried out at the *Institute of Micro- and Nanoelectronic Systems* (IMS) at the *Karlsruhe Institute of Technology* (KIT), Karlsruhe, Germany. However, specific experiments were carried out at collaborating Institutes at the KIT, the *University of Tübingen*, Germany, the *Technical University of Denmark* and the *Russian Academy of Science*. The help and assistance I have received from many friends and colleagues has made this work possible.

Citations

For clarification, three different types of references have been used within this thesis, presented in distinct styles.

- External publications are numbered after their occurrence in the text.

- My co-authored publications are cited by the first letters of the first 3 - 4 authors' surnames, followed by the year of publication. If there are more than 4 authors, it is indicated by a "+" sign.

- Student theses that I have supervised throughout this work are cited by the first three letters of the students' surnames and the year of completion of the theses.

Complete lists of my own publications, supervised student theses, as well as a chronological list of all international conferences at which my findings have been presented, are given towards the end of this thesis.

Karlsruhe,
August 2013

Johannes Maximilian Meckbach
Karlruhe Institute of Technology (KIT)

i

Danksagung

An erster Stelle möchte ich mich bei Prof. Michael Siegel dafür bedanken, dass er mir ermöglicht hat am Institut für Mikro- und Nanoelektronische Systeme (IMS) zu promovieren.
Stets hat er mir Freiraum gelassen meine Forschung eigenständig zu gestalten und mich in
meinen wissenschaftlichen Vorhaben unterstützt. Die Mittel bewilligt zu bekommen eine
komplette Sputteranlage neu konzipieren und bauen zu dürfen ist nur ein Beispiel für das
Vertrauen, das er mir entgegengebracht hat. Natürlich möchte ich mich auch für die Mög-
lichkeit bedanken, dass ich meine Ergebnisse auf diversen Konferenzen präsentieren durfte.

Bei Prof. Reinhold Kleiner möchte ich mich selbstverständlich ebenfalls herzlichst bedanken. Zum einen für die Übernahme des Korreferats und zum anderen für seine ständige
sowohl fachliche als auch menschliche Unterstützung über die gesamte Dauer meiner Promotion hinweg. Die vielen Besuche in Tübingen und Treffen auf diversen Konferenzen
haben mir immer viel Freude bereitet. Ich hoffe wir werden weiterhin in Kontakt bleiben.

Auch bei Prof. Dieter Kölle möchte ich mich herzlichst bedanken. Egal ob wir uns in
Tübingen, auf Konferenzen oder auf Feiern getroffen haben, hatte ich immer Spaß und
konnte viel lernen. Die zahlreichen Diskussionen über SQUIDs, κ-Fluxonen oder Mess-
systeme haben mir viel fachliches Verständnis gebracht. Danke dafür, dass ich mich mit
jeglichen Problemen immer melden konnte. Ebenfalls aus der Tübinger Gruppe möchte ich
mich bei Dr. Edward Goldobin bedanken. Seine Geduld bei den vielen Designiterationen
für κ-Fluxon Devices war für mich unbezahlbar und hat meine Promotion in großen Teilen
erst ermöglicht. Dass er sich immer Zeit genommen hat mir jedes kleine Detail zu erklären
ist bei Weitem nicht selbstverständlich. Ich würde mich freuen auch in Zukunft weiter neue
Projekte mit Dir durchzuführen.

Ein riesen Dank geht natürlich auch an alle Mitarbeiter des IMS. Zunächst Dr. Konstan-
tin Ilin, der mir in der Technologie immer wieder gute Ratschläge gegeben hat und mich
oft beim Bau der neuen Sputteranlage unterstützt hat. Dr. Christoph Kaiser, dessen Erbe ich
am IMS angetreten habe möchte ich natürlich ebenso für die lehrreiche Anfangszeit danken. Vielerlei Unterstützung habe ich auch von Dr. Stefan Wünsch bei jeglichen Fragen zu
Hochfrequenzangelegenheiten erhalten, wofür ich auch Ihm zu Dank verpflichtet bin. Natürlich danke ich auch meinen ursprünglichen Zimmerkollegen Alexander Scheuring und Axel
Stockhausen für eine gute Zeit im Raum 126. Petra Thoma möchte ich noch ganz besonders
danken. Die immerwährende Aufmunterung von Ihr hat mir jeden Arbeitstag versüßt. Irgendwann gibt es sicher „Pro-Island". Matthias Hofherr danke ich für die Unterstützung bei
der Suche diverser Masseschleifen und für die unfassbar leckeren süßen Stückchen. Dagmar

Henrich und Gerd Hammer als Vertreter der „alten Genration" und Matthias Arndt, Artyom Kuzmin, Juliane Raasch und Philipp Trojan als Vertreter der „neuen Generation" danke ich für viele spaßige Momente.

Besonderen Dank möchte ich meinem „JJ-Nachfolger" Michael Merker aussprechen, mit dem ich viel Freude in der Technologie, im Reinraum, im Büro und auch außerhalb des IMS hatte. Der mit Dir zusammen entwickelte Prozess war ein Meilenstein meiner Promotion und ich danke Dir für Deine Hilfe. In diesem Zuge möchte ich mich auch noch bei den von mir betreuten Studenten, Simon Bühler, Wolfgang Heni, Sebastian Heunisch, Alexander Lochbaum und Pavol Marko bedanken. Besonders hervorheben möchte ich die herausragende Mitarbeit von Simon, Pavol, Alex und Michi (während seines Studiums). Egal ob spät Nachts, am Wochenende oder am Feiertag, waren sie immer bereit auch weit mehr als nötig gewesen wäre zu erledigen. Ohne Euch hätten es niemals so gute JJs am IMS gegeben und meine Zeit am IMS wäre nicht annähernd so witzig geworden.

Weiter will ich noch die großartige Zusammenarbeit mit Dr. Joachim Nagel, Matthias Rudolph, Uta Kienzle, Dr. Kai Buckenmaier, Dr. Tobi Gaber der Universität Tübingen erwähnen. Ohne sie hätte wohl kaum jemand Interesse an meinen JJs gehabt. Danke dafür, dass ich zu jeder Zeit (auch gerne mal am Wochenende) bei Euch messen konnte und wir zum Teil auch schöne Veröffentlichungen realisiert haben. Die legendären Abende im Besen werde ich nie vergessen. Besonders „MessMatze" wünsche ich noch eine weiterhin erfolgreiche Promotion. Ich bin mir sicher die Zusammenarbeit mit Michi wird super. Danke an Euch Alle noch dafür, dass ich immer bei einem von Euch übernachten konnte.

Einen entscheidenden Anteil an dieser Arbeit hatten noch Alexander Stassen, Hansjürgen Wermund und Karlheinz Gutbrod. Dabei gilt mein Dank in besonderem Maße Karlheinz Gutbrod und Alexander Stassen, die zusammen mit mir die neue Sputteranlage zum größten Teil gebaut haben. Ohne sie wäre ich kläglich gescheitert. Ich hoffe ich kann mich irgendwann noch für die ständige Hilfe revanchieren. Weiter will ich mich noch bei Frank Ruhnau für die IT-Unterstützung, Erich Crocoll für die Unterstützung im täglichen Ablauf am IMS und Doris Duffner für die Abwicklung aller bürokratischen Aspekte meiner Promotion bedanken.

Andreas Häffelin danke ich für die vielen XRD Messungen meiner NbN und AlN Proben. Ohne Dein Engagement hätte ich niemals eine solche Qualität an NbN/AlN/NbN JJs erreicht. Hoffentlich sehen wir uns bald mal wieder. Martin Weides, Jochen Braumüller und Sebastian Skacel danke ich für die Unterstützung bezüglich des Atomic Force Microscope. Durch Euch konnte ich überhaupt erst die richtigen Parameter für glatte NbN Elektroden finden.

Abschließend möchte ich noch einigen Menschen aus meinem privaten Umfeld danken. Mit meinen Eltern und meiner Schwester habe ich einfach das große Los gezogen. Ihr seid unfassbar für mich. (Ich hoffe meine Arbeit hilft wenn ihr versucht meine kleine Nichte, Clara, zum Einschlafen zu bringen.) Pat, Simon, Marc, Raphi, Conny und Katha: Ich schätze mich glücklich Euch meine Freunde nennen zu können. Egal ob auf Schnee, Wasser oder Asphalt hatte ich immer eine tolle Zeit mit Euch und ihr seid stets der Gegenpol zu den langen Nächten im Reinraum und der Technologie gewesen. Auch möchte ich Rosie und Uli danken, die mir mit dem englischen Text sehr geholfen haben. Unmöglich wäre aber all dies hier gewesen, wenn nicht Silvester 2010 die tollste Frau überhaupt in mein Leben getreten wäre. Lisa, Du machst mich so glücklich wie niemand sonst.

Kurzfassung

Im Zuge der stetigen technologischen Weiterentwicklung müssen immerfort neue Technologien entwickelt werden um die bestehenden Grenzen zu überwinden. In den vergangenen Jahrzehnten wurden dafür zunehmend gekühlte Systeme, basierend auf Supraleitern, eingesetzt. Josephson-Kontakte bilden hierbei grundlegende Elemente zur Untersuchung fundamentaler physikalischer Effekte, sowie für die Realisierung von ultra-schneller Elektronik und hochsensibler Detektoren, deren Leistungsfähigkeit die der konventionellen Halbleitertechnologie weit übertrifft. Im Rahmen dieser Arbeit wurden technologische Herstellungsprozesse für Josephson-Kontakte entwickelt und grundlegend erweitert. Dies umfasst sowohl den weltweit standardmäßig verwendeten Prozess mit der Schichtabfolge Niob / Aluminiumoxid / Niob (Nb/Al-AlO$_x$/Nb), als auch die Neuentwicklung eines Prozesses basierend auf Niobnitrid / Aluminiumnitrid / Niobnitrid (NbN/AlN/NbN). Mit dem Nb/Al-AlO$_x$/Nb Prozess wurden in der vorliegenden Arbeit diverse Bauelemente entwickelt und eine Vielzahl von Experimenten daran durchgeführt, wobei eine Verbesserung aller untersuchten Devices gezeigt werden konnte. Hinsichtlich zukünftiger Josephson-Bauelemente basierend auf NbN/AlN/NbN war es möglich, eine signifikante Verbesserung gegenüber der herkömmlichen Schichtabfolge zu erzielen.

Seit ihrer theoretischen Vorhersage durch Brian David Josephson in Jahr 1961, wurden viele unterschiedliche Bauweisen für Josephson-Kontakte realisiert. Der mit Abstand am weitesten verbreitete Prozess basiert auf einem Dreilagensystem aus zwei supraleitenden Elektroden, die durch eine sehr dünne, elektrisch isolierende Schicht, voneinander getrennt sind. Typischerweise werden solche Kontakte auch als Supraleiter-Isolator-Supraleiter oder SIS-Kontakte beschrieben. In einem solchen System kann es bei einer hinreichend dünnen Barriere zu einer schwachen Kopplung der makroskopischen Wellenfunktionen der beiden Supraleitender kommen. Infolgedessen können Ladungsträger durch diese Potentialbarriere tunneln. Ein Josephson-Kontakt entspricht folglich einem Tunnelkontakt, bei dem makroskopische Größen, der Tunnelstrom und die Spannung über den Kontakt, direkt von einer quantenmechanischen Größe, der Phasendifferenz der Wellenfunktionen, abhängen. Josephson-Kontakte finden heutzutage in vielen Bauelementen Anwendung. Die prominenteste Anwendung ist wahrscheinlich der Josephson-Spannungsstandard, der seit 1990 die Einheit *Volt* definiert. Weiterhin werden Josephson-Kontakte auch in ultra-schneller Elektronik, sogenannter Rapid Single-Flux Quantum (RSFQ) Schaltungen, in Detektorsystemen für hochsensible Magnetfeldmessungen, oder Hochfrequenzstrahlung eingesetzt. Nicht zuletzt sind sie aber durch ihre einzigartige Verbindung der makroskopischen mit der quantenmechanischen Physik auch von großem Interesse für fundamentale physikalische Experimente.

Die Zielsetzung dieser Arbeit war es, den bestehenden konventionellen Nb/Al-AlO$_x$/Nb Prozess so zu erweitern, dass komplexe Bauelemente mit sub-µm Abmessungen in allen Lagen herstellbar sind. Zusätzlich war es das Ziel, Josephson-Bauelemente weiterzuentwickeln und somit deren Funktionsweise zu verbessern. Anhand der gewonnenen Erkenntnisse sollten Anforderungen an neue Fabrikationstechnologien definiert werden und ein neuer Herstellungsprozess für zukünftige Bauelemente entwickelt werden. Diese Neuerungen haben sowohl die Prozessabfolge, als auch die verwendeten Materialien betroffen. Die Neuerungen in der Prozessabfolge sind umfassend in Abschnitt 3.1 und 3.2 diskutiert. Sowohl der Prozess basierend auf den neuen Materialien Niobnitrid und Aluminiumnitrid, als auch die dazu entwickelte neu aufgebaute Depositionsanlage sind in Kapitel 4 zu finden. Zur Umsetzung dieser Ziele war es notwendig das Bedienen einer Vielzahl von Geräten zu erlernen. Neben UV- und Elektronenstrahllithographie, beinhaltete dies auch verschiedene Abscheide- und Ätzprozesse. Die Qualität der gefertigten Proben wurde stetig mittels optischer Mikroskopie, Elektronenstrahlmikroskopie und Transportmessungen untersucht.

Die Qualität von Josephson-Kontakten wird typischer Weise, anhand von Gleichstrommessungen ermittelt. Aus Strom-Spannung Kennlinien können diverse Qualitätsparameter extrahiert werden, die auf die Qualität der supraleitenden Elektroden, das Tunnelverhalten und die Dichte der Tunnelbarriere schließen lassen. Zusätzliche Messungen der Kontakte in einem extern angelegten Magnetfeld ermöglichen Rückschlüsse auf die Homogenität der Stromdichterverteilung innerhalb des Kontakts. Solche Messungen können auf Kontakte kleiner Abmessungen ohne Probleme angewendet werden. Übertreffen die Abmessungen jedoch eine gewisse charakteristische Größe, die zusätzlich von der kritischen Stromdichte der SIS Multilage abhängig ist, so treten Inhomogenitätseffekte auf, die keine eindeutigen Rückschlüsse auf die Kontaktqualität mehr erlauben. Da viele der durchzuführenden Experimente auf langen Josephson-Kontakten basierten, musste daher zunächst eine verlässliche Methode entwickelt werden, mittels derer die Kontaktqualität trotzdem ermittelt werden kann. Hierzu wurden Untersuchungen zunächst an kurzen Kontakten und dann an Kontakten im Größenbereich der charakteristischen Abmessung durchgeführt. Um auch lange Kontakte anhand klassischer Strom-Spannungs-Kennlinien eindeutig charakterisieren zu können, wurden neue Zuleitungen entwickelt, die eine homogene Stromeinspeisung ermöglichen. Nachdem somit eine hohe Qualität nicht nur kurzer, sondern auch langer Kontakte nachweisbar war, wurden weiterführende Experimente durchgeführt.

Das erste thematisierte Bauelement in dieser Arbeit ist ein Superconducting Quantum Interference Device (SQUID), das hinsichtlich seiner Energieauflösung optimiert wurde. Hierzu ist es notwendig die Spannungsantwort des SQUIDs auf Änderungen im externen, zu detektierenden magnetischen Fluss zu maximieren, ohne dabei das Rauschen des SQUIDs im selben Maße zu erhöhen. Dieser Teil der Arbeit wurde in Zusammenarbeit mit dem *In-*

stitut für Experimentalphysik II (PIT II), der *Universität Tübingen* durchgeführt. Bei dem Design und der Herstellung lag die Herausforderung darin, die parasitären Einflüsse des angelegten Magnetfelds auf das Bauelement maximal zu unterdrücken, ohne dass die Funktionalität darunter leidet. Die Designmaßnahmen und die durchgeführten Experimente sind in Abschnitt 2.4.1 dargelegt. Mittels der durchgeführten Neuerungen, war es möglich, die normierte Energieauflösung um den Faktor 3,4 in Relation zu dem herkömmlichen Bauelement zu verbessern. Solche verbesserten SQUIDs bergen enormes Potential für optimierte supraleitende Verstärker oder Magnetfeldsensoren.

Im weiteren Verlauf dieser Dissertation werden Bauelemente für fundamentale Untersuchungen fraktionaler magnetischer Flusswirbel, basierend auf langen Josephson-Kontakten diskutiert. Auch dieser Teil der Arbeit wurde in enger Zusammenarbeit mit dem PIT II, der *Universität Tübingen* durchgeführt. Stark unterdämpfte lange Kontakte stellen eindimensionale Systeme für magnetische Flusswirbel dar. Diese Flusswirbel, auch Josephson Vortizes, oder Fluxonen genannt, können sich entlang eines langen Kontakts frei bewegen und sind solitonischer Natur. Natürlich auftretende Fluxonen tragen stets ein magnetisches Flussquant $\Phi_0 = 2,07 \cdot 10^{-15}\,\text{Wb}$. Diese Quantisierung kann mittels zusätzlicher, extrem kleiner DC Zuleitungen in einer der supraleitenden Elektroden gebrochen werden. Somit können fraktionale Vortizes realisiert werden. Die *Größe* dieser Vortizes kann durch einen DC Strom beliebig zwischen 0 und Φ_0 eingestellt werden. Solche fraktionale Vortizes sind an ihren Entstehungsort gebunden, sind also keine Solitonen und ihnen kann eine äquivalente Masse und topologische Ladung zugeordnet werden. Dadurch verhalten sie sich wie isolierte einzelne Teilchen in einem eindimensionalen System, weisen eine Eigenfrequenz auf und können sich im Falle mehrerer Vortizes innerhalb eines Kontaktes gegenseitig beeinflussen. Einige dieser Experimente machten eine grundlegende Erweiterung des Herstellungsprozesses notwendig, um in allen Lagen der Multischicht-Bauelemente laterale Abmessungen im sub-µm Bereich zu erzielen. Hierzu wurde ein neuer Prozess entwickelt, der eine planare Chiptopographie erzeugt. Die minimal erzielte Josephson-Kontakt Größe konte auf einen Durchmesser von $d_{\text{JJ}} \approx 200\,\text{nm}$ reduziert werden und zusätzliche Metallisierungsebenen konnten eingebunden werden. Anhand unterschiedlicher Kontakte annularer oder linearer Geometrie, wurde die Dynamik solcher fraktionalen Vortizes untersucht. Erstmalig konnte im Rahmen dieser Arbeit anhand spektroskopischer Messungen die Abhängigkeit der gegenseitigen Beeinflussung fraktionaler Flussquanten von ihrem Abstand zueinander gezeigt werden. Zusätzlich wurden verschiedene Zustände linearer Vortexmoleküle mittels strom-asymmetrischer SQUIDs bei unterschiedlichen Temperaturen bis in den mK-Bereich untersucht. Basierend auf solchen Bauelementen sind neuartige Bauelemente, wie zum Beispiel einstellbare Hochfrequenzfilter, Metamaterialien oder Quantenbits denkbar. Die Charakterisierung solcher fraktionaler Josephson Vortizes werden in Abschnitt 3.4.2 vorgestellt und die Resultate der weiterführenden Untersuchungen werden in Abschnitt 3.4.3 und 3.4.4 diskutiert.

Als drittes Bauelement wurden Lokaloszillatoren, sogenannte flux-flow Oszillatoren (FFOs), untersucht. Legt man ein hinreichend großes Magnetfeld an einen langen Josephson-Kontakt an, bilden sich magnetische Flusswirbel über die Barriere hinweg aus, die mittels eines angelegten DC Stromes und der daraus resultierenden Lorentzkraft beschleunigt werden können. Am Ende des Kontakts tritt folglich eine hochfrequente Änderung der Phasendifferenz auf, was zur Emission elektro-magnetischer Strahlung führt. Die Frequenzstabilität dieser Strahlung definiert die Linienbreite der Emission und ist für die Verwendung des Kontakts als Lokaloszillator im sub-THz Bereich von zentraler Bedeutung. Bei der Weiterentwicklung solcher FFOs, lag ein Hauptaugenmerk auf der Reduzierung dieser Linienbreite mittels einer künstlichen Korrelation des Stromes zur Erzeugung des Magnetfeldes und des Stromes zur Erzeugung der Lorentzkraft. Zusätzlich konnte anhand der Messungen ein neues Simulationsmodell von Dr. Sobolev vom *Institute of Radioengineering and Electronics* (IRE) an der *Russian Academy of Sciences* (RAS) in Moskau, Russland, für die Berechnung von Mikrowellenleitungen verifiziert werden. Die Messungen der Linienbreiten bei Frequenzen bis zu $f_{FFO} \approx 360 \, \mathrm{GHz}$ mittels eines integrierten SIS-Frequenzmischer-Elements wurden an dem *Physikalischen Institut* der *Technischen Universität Dänemarks* in Lyngby, Dänemark, durchgeführt. Die in Abschnitt 3.4.5 gezeigten Experimente zeigen, dass eine leichte Reduzierung durch einen zum FFO parallelgeschaltet, seriellen *LC*-Schwingkreis, erzielt werden können und die Mikrowellentransmission anhand des neuen Modells sehr gut vorhergesagt werden können.

Der letzte Teil dieser Arbeit befasst sich mit der Entwicklung einer neuen Multilagen-Technologie für verbesserte Josephson-Bauelemente. Im Unterschied zu den zuvor besprochenen Bauelementen, sollten diese auf der Multilage NbN/AlN/NbN basieren. Die Verwendung von Niobnitrid anstatt von Niob bringt eine erhöhte Grenzfrequenz für Strahlungsdetektoren und Strahlungsemitter mit sich. Hiermit können Elemente für den THz-Frequenzbereich realisiert werden, der bis vor wenigen Jahren noch nahezu unerforscht blieb. Des Weiteren fällt der Entwurf neuer Bauelemente durch die Verwendung von Aluminiumnitrid anstatt von Aluminiumoxid als Tunnelbarriere, aufgrund der geringeren spezifischen Kapazität der neuen Schichtabfolge, deutlich leichter aus. Für die Neuentwicklung dieser Technologie wurde zunächst eine Dreikammer-Sputteranlage für epitaktische Filmabscheidung von Grund auf neu konzipiert. Die Anlage umfasst sowohl eine *in-situ* Reinigung der Substrate, zwei 3-Zoll DC Magnetrons zum reaktiven Abscheiden von NbN und AlN, als auch einen beweglichen Heiztisch. Die sorgfältige Optimierung der einzelnen Schichten ermöglichte, dass epitaktische NbN und AlN Schichten auf unterschiedlichen Substraten abgeschieden werden können und enorm hohe kritische Temperaturen von NbN auf Siliziumsubstraten mit Hilfe eine AlN-Pufferschicht erreicht wurden. Diese liefern die Basis zur Herstellung von qualitativ sehr guten NbN/AlN/NbN Josephson-Kontakten. Zur vollständigen Charakterisierung der NbN/AlN/NbN Multilagen ist es notwendig Josephson-Kontakte zu vermessen. Da die vollständige Fertigung jedoch über eine Woche in Anspruch nimmt, ist eine umfassende Charakterisierung sehr zeitaufwendig. Hinsichtlich einer zeit-optimierten Fertigung von Josephson-Kontakten wurde daher ein weiterer Herstellungs-prozess neu entwickelt, der es ermöglicht, innerhalb von lediglich drei Tagen vollständige Josephson-Bauelemente herzustellen und somit eine schnelle und komplette Charakterisie-rung neuer Multilagen ermöglicht. In Kapitel 4 wird die neue Mehrkammer-Sputteranlage vorgestellt, die Resultate der individuellen Schichtoptimierung diskutiert und anschließend werden die erzielten Qualitätsparameter der NbN/AlN/NbN Josephson-Kontakte auf unter-schiedlichen Substratmaterialien präsentiert.

Abschließend kann zusammengefasst werden, dass im Rahmen dieser Arbeit umfassende Erweiterungen des konventionellen Nb/Al-AlO$_x$/Nb Prozesses zu einer enormen Miniaturi-sierung der lateralen Abmessungen in allen Schichten des Multilagenprozesses in den sub-µm Bereich geführt hat. Zugleich, konnte eine hohe Qualität von kurzen und auch langen Josephson-Kontakten erreicht werden. Basierend auf diesen Untersuchungen, wurden wei-terführende Experimente an SQUIDs, fraktionalen Vortex Bauelementen und flux-flow Os-zillatoren durchgeführt und Verbesserungen aller Bauelemente erzielt. Zusätzlich wurde ein vollständig neuer Fabrikationsprozess für die Herstellung von NbN/AlN/NbN Josephson-Kontakten entwickelt. Die neuen Kontakte bringen diverse Vorteile für den Entwurf von Hochfrequenzbauelementen und weisen sehr gute Qualitätsparameter auf.

Contents

1. Introduction and Motivation

In 1961 Brian D. Josephson predicted that two weakly coupled superconducting wavefunctions Ψ_1 and Ψ_2 can result in a non-dissipative tunneling of superconducting charge carriers, so-called Cooper-Pairs, through a properly designed potential barrier [1]. This behavior is well described by the first Josephson equation, which links the direct current (dc) through such a Josephson junction (JJ) to the difference between the phases φ_{sc1} and φ_{sc2} of the two macroscopic wavefunctions. Above a certain critical value of the dc current applied to the junction, the charge transport is no longer non-dissipative, resulting in a discontinuous jump of the voltage, which is dependent on the change of the phase difference over time and possible magnetic fields penetrating the structure. Such a Josephson junction represents a system in which the two macroscopic measures, the dc current and voltage, are directly linked to a quantum measure, the phase difference of the superconducting wavefunctions. JJs are typically categorized in short and long Josephson junctions and are used in both, experiments involving classic physics as well as quantum experiments.

Today, Josephson junctions are utilized in many applications including the voltage standard [2–6], ultra-fast electronics (so-called Rapid Single-Flux Quantum, or RSFQ) [7], local oscillators for the generation of radiation up to the THz range [8–10], high frequency receiver devices [11, 12] and the most sensitive detector elements for magnetic fields, so-called Superconducting Quantum Inference Devices (SQUIDs) [13, 14]. Although highly interesting from the point of view of application, some of these devices, such as the voltage standard or RSFQ logic, are almost impossible to realize using fabrication facilities available in a university laboratory, since they may require a large number of JJs and thus quickly reach high levels of complexity [15]. In contrast, devices for the emission and reception of electromagnetic radiation, SQUIDs or devices particularly designed for the investigation of fundamental physics, typically require just a few JJs on each chip. Focusing on such applications, fabrication technologies for various different Josephson junction devices have been developed within this thesis. The investigated devices, fabricated using suitable processes, as well as the results obtained from the performed experiments are discussed in this thesis.

In chapter 2 some fundamental aspects of superconductivity are introduced, forming a sufficient basis for the introduction of the Josephson equations before the characterization procedure for JJs is presented. Afterwards, the investigated Josephson devices are discussed in more detail, outlining the limitations of their performance and ways of overcoming these limitations to some degree. In some cases this can be done by means of design aspects, whereas some limitations require an alteration of the fabrication process or material composition. At the beginning of chapter 3 the conventional fabrication process is introduced, which was already available prior to this thesis. Before the specific devices developed within the framework of this thesis are discussed, the newly developed refined fabrication process is introduced. This new fabrication process allows sub-µm patterning in all layers as well as the integration of higher metallization levels due to a planar chip topography.

With the focus on enhancing the performance of dc SQUIDs, a new way of resistively shunting the junctions has been investigated. Preliminary numerical simulations, based on the Langevin equations, were performed at the *University of Tübingen*. The simulations predicted that the energy resolution, defined by the maximum voltage response per applied magnetic flux with respect to the noise afflicted with the device, can be optimized by creating a strong asymmetry in the shunt resistance. This asymmetry was realized by removing one of the resistors and replacing the other with half of its original resistance. Here, the technological challenge was the minimization of parasitic resonances, by a proper design and fabrication of the shunt resistor loop. The resulting device has been compared to its fully symmetric counterpart and a performance enhancement in terms of the normalized energy resolution by a factor of 3.4 was found.

Using long Josephson junctions (LJJs), fundamental physics involving vortex dynamics has been investigated. In such LJJs magnetic fields may penetrate in the form of so-called Josephson vortices, or fluxons, carrying one magnetic flux quantum Φ_0. These fluxons may move freely along the LJJ, which forms a one-dimensional system, and thus exhibit solitonic behavior. By means of special designs, it is possible to create vortices carrying any arbitrary fraction of Φ_0 and to pin them to their point of origin [16, 17]. Such fractional vortices behave like individual particles and measures such as mass and a topological charge may be ascribed to them. Systems incorporating one or two fractional vortices have been investigated in linear and annular LJJs with respect to their behavior under external influences, such as microwave radiation, bias currents or variations in temperatures. Due to the challenging design and fabrication, a set of requirements has been defined for a more versatile and sophisticated fabrication process, previously introduced as the refined process. Technologically, long JJs for the investigation of fractional vortices, are extremely challenging, due to the necessity of sub-µm patterning of almost all layers.

Another device based on fluxon dynamics is the so-called flux-flow oscillator (FFO). This device was first proposed by Nagatsuma *et al.* to serve as a local oscillator integrated on-chip, emitting frequencies up to the gap-frequency of the superconducting electrodes [8]. For the widely used niobium - aluminum oxide - niobium multilayer process, this frequency reaches up to $f_g \approx 677\,\text{GHz}$. When integrated on-chip together with a receiver device, *e.g.* a very small JJ, the performance of the resulting receiver chip is strongly influenced by the emission linewidth of the FFO. For optimal performance of such devices, the critical-current density had to be increased significantly. Within this thesis improved high-j_c FFOs have been developed, which include a feedback circuit, reducing the fluctuations of the emission and thus reducing the linewidth. The results achieved are presented at the end of chapter 3.

Many Josephson devices would benefit from a multilayer technology, based on superconducting materials with a higher energy gap than that of niobium. This would directly result in a larger voltage jump upon crossing the critical value of the bias current, above which the charge transport becomes dissipative. Since this gap voltage directly corresponds to a gap-frequency, devices suitable for higher frequencies could be realized. Furthermore, the design restrictions for high frequency devices and SQUIDs are relaxed somewhat with a decreasing specific capacitance of the Josephson junctions. Being dominated by the insulating material, the specific capacitance can be reduced by choosing a material with a smaller effective dielectric constant than that of the commonly used aluminum oxide. For enhancement of future devices a multilayer process based on niobium nitride and aluminum nitride has thus been developed. These materials fulfill the above mentioned requirements and are thus ideal for devices operating in the THz range. For this, a completely new three-chamber sputter system, with an *in-situ* pre-cleaning, two dc magnetrons and a movable heater has been designed and constructed within the scope of this thesis, which allows epitaxial growth of NbN and AlN layers. Thin films for various detector systems based on NbN as well as trilayers of NbN/AlN/NbN for Josephson junctions have been developed and investigated. The new deposition procedures have been optimized on different substrates including magnesium oxide and sapphire. Particular focus has been laid on high-resistive silicon as a substrate material, being transparent for infrared, as well as THz radiation, while still being low cost and allowing fast implementation into existing fabrication processes. The results obtained with this new fabrication process based on niobium nitride and aluminum nitride, are presented in chapter 4.

2. Superconductivity and Josephson Effects

In this chapter a short history of the discovery of superconductivity and the phenomena thereof will be introduced briefly in section one. Afterwards, dc and ac-Josephson effects relevant to this work, as well as different activation mechanisms for short junctions will be discussed. The third section deals with the dynamics of long Josephson junctions and the various excitations of the Josephson phase will be explained. The last section introduces the devices investigated and explains the reasons for the technological developments realized as part of this research.

2.1. Fundamentals of Superconductivity

In the early 1900s, before Heike Kamerlingh Onnes discovered superconductivity, there was a strong international interest in reaching lower temperatures. One of the main reasons was the investigation of the electrical resistance R of metals at temperatures close to absolute zero, since there were three competing opinions on possible properties. James Dewar proposed that the resistance of pure metals would approach $R = 0\,\Omega$ with decreasing temperature, while Heinrich F. L. Matthiesen suggested that the resistance could remain on a remanent plateau, with $R > 0\,\Omega$, below a material dependent temperature. Onnes' lab books indicate that he believed in the third opinion, suggested by William Lord Kelvin in 1902, stating that the mobility of the electrons would vanish with decreasing temperature, resulting in an exponential increase of the resistance [18]. In 1908, Onnes succeeded in liquefying helium and achieved a new record in low temperatures. Even though in 1910, $R(T)$ measurements of a platinum wire proved Matthiesen's model to be correct, further measurements of different metals were continued. On April 8th, 1911, Onnes measured the resistance of mercury and found a sudden drop to immeasurably small values for $T \leq 4.19\,\mathrm{K}$ and currents below a *critical-current density* $j_{\mathrm{s,c}}$ [18]. Such transition temperatures, at which the dc resistance of a material drops to zero, is called *critical temperature* T_{c} and the state to which the material passes, is called *superconductivity*. Ever since 1911, thousands of superconducting materials, elements as well as compounds, have been found [19]. In 1913, Heike Kamerlingh Onnes received the Nobel Prize in physics for this discovery.

Despite the fact that this discovery sparked enormous interest amongst scientists worldwide, the phenomenon could not be explained until many years later. After the discovery of this perfect conductivity, proven by persistent currents in superconducting rings, it took Walther Meissner and Robert Ochsenfeld until 1933 to discover the diamagnetic properties of superconductors. They found that magnetic fields are not only excluded from entering a superconductor, but also expelled when being cooled below T_c, except for a thin layer on the surface of the superconductor. This perfect diamagnetism inside bulk superconductors proved superconductivity to be a new, independent thermodynamic state. The maximum magnetic field above which the superconductivity collapses is called *critical field* H_c and can be approximated by $H_c(T) \approx H_c(0) \cdot [1 - (T/T_c)^2]$, where T denotes the temperature. Besides the critical temperature T_c and the critical-current density $j_{s,c}$, the critical magnetic field H_c is the third material dependent parameter, below which the superconducting state persists.

In 1935 the brothers Fritz and Heinz London proposed two equations [20]:

$$\vec{E} = \mu_0 \lambda_L^2 \frac{\partial}{\partial t} \vec{j}_s \tag{2.1}$$

$$-\vec{B} = \mu_0 \lambda_L^2 \vec{\nabla} \times \vec{j}_s \tag{2.2}$$

describing the microscopic electric and magnetic field, respectively. Here μ_0 is the vacuum permeability and j_s denotes the supercurrent density.

$$\lambda_L = \sqrt{\frac{m_s}{\mu_0 q^2 n_s}} \tag{2.3}$$

is the *London penetration depth* and corresponds to the length scale on which magnetic fields exponentially decay within a superconductor. m_s is the mass, q_s the charge and n_s the density of the superconducting charge carriers. Consequently, λ_L is the thin surface layer in a superconducting material where currents may flow, screening external magnetic fields and hence giving rise to the perfectly diamagnetic core of a bulk superconductor. The technological importance of λ_L will be discussed in more detail in subsection 3.4.2.

In 1957, Bardeen, Cooper and Schrieffer published the first microscopic theory of superconductivity, the so-called *BCS theory* [21]. The theory's hypothesis was based on the possibility of an attractive force between two electrons, first presented by Herbert Fröhlich in 1950 [22]. Fröhlich proposed that two electrons may be attracted to each other, despite their equally negative charge, by interaction through lattice vibrations. The BCS theory showed that this force allows two electrons to enter a bound state, known as Cooper-pairs, named after Leon Neil Cooper [23]. Upon entering this bound state, the two electrons relax into a more energy efficient state, as compared to their unbound state, which was first shown by measurements of the specific heat of superconductors [24, 25], *i.e.* when entering the superconducting state, an energy gap $\Delta(T)$ around the Fermi level is formed. For $T < T_c$, Cooper-pairs occupy the Fermi level and may only be broken up into two single electrons, called *quasiparticles*, when an energy $E \geq 2\Delta(T)$ is applied. The theory predicted, that the energy gap rises from zero at $T = T_c$ to

$$\Delta(0) = 1.764 k_B T_c, \tag{2.4}$$

at $T = 0\,\text{K}$, where k_B is the Boltzmann constant.

Cooper pairs comprise of two electrons, having equal and opposite spins and momentum. All charge carriers within one superconductor may thus be described by a single macroscopic wavefunction

$$\Psi(\vec{z},t) = \sqrt{n_s} \cdot e^{i\varphi_{sc}(\vec{z},t)} = |\Psi(\vec{z},t)| \cdot e^{i\varphi_{sc}(\vec{z},t)}. \tag{2.5}$$

Here φ_{sc} is the phase and the Cooper pair density $n_s = |\Psi|^2$ is the amplitude. $\vec{z}$ and t denote the spatial coordinate and time, respectively. With the size of the superconductor being arbitrary, Ψ describes a quantum phenomenon on macroscopic scale. The fact that Ψ is phase coherent is due to the size of a Cooper pair, which can reach up to a few µm and is therefore much larger than the distance between individual Cooper-pairs which may of course overlap. This size is at the order of the BCS coherence length ξ_0. Already been published in 1950, 7 years prior to the BCS theory, the Ginzburg-Landau theory had published [26]. The GL theory introduced the so-called the Ginzburg-Landau coherence length $\xi(T)$, which is typically in the nm range and describes the length scale on which Ψ may vary. At this point it should be mentioned that besides $\xi(T)$, the GL theory also introduced a complex order parameter $|\Psi(z)|^2 = n_s$, equivalent to the macroscopic wavefunction in Eq. (2.5). The GL theory furthermore introduced a parameter

$$\kappa_{GL} = \frac{\lambda_L}{\xi}. \tag{2.6}$$

For typical superconductors investigated during that time period, $\lambda_L \approx 50\,\text{nm}$ and $\xi \approx 300\,\text{nm}$ and hence $\kappa_{GL} \gg 1$ [27]. Abrikosov however, investigated the case of $\kappa_{GL} < 1$ and found that for $\kappa_{GL} < 1/\sqrt{2}$ the breakdown of superconductivity, due to an externally applied magnetic field, is not discontinuous anymore, but that flux may penetrate the superconductor once H exceeds a first lower critical field $H_{c,1}$ [28]. Only at a higher value $H_{c,2}$ is the material fully penetrated. He called these materials type-II superconductors, whereas materials with $\kappa_{GL} \geq 1/\sqrt{2}$ are called type-I superconductors. Furthermore, he found that the flux penetrated in quantized tubes of magnitude

$$\Phi_0 = \frac{h}{2e} = 2.07 \times 10^{-15}\,\text{Wb}, \tag{2.7}$$

having a normal core with $|\Psi|^2 = 0$. Here e is the charge of a single electron and h is Planck's constant. This quantized flux penetration is comparable to the flux quantization in superconducting rings. When taking a closer look at Eq. (2.5), it becomes obvious that the wavefunction must be 2π-periodic in a ring structure. External magnetic fields however, may be of any arbitrary magnitude and hence could induce circulating currents, such that φ is not 2π-periodic. A wave traveling around the ring would destructively interfere with itself. Evidently, this is impossible, which was first addressed by F. London in 1950 [29]. To compensate, a circulating supercurrent is induced, such that the phase is again 2π-periodic. In turn, this results in a quantization of the magnetic flux through the area enclosed by the superconducting ring. Experimental demonstration of this effect followed in 1961 by B. S. Deaver and W. M. Fairbank in Stanford [30] and R. Doll and M. Näbauer in Munich [31].

Conclusion

Here, fundamental effects of superconductivity have been introduced, forming a sufficient base for further reading of this thesis. Additionally, a brief outline of the history of superconductivity was given.

2.2. Short Josephson Junctions

Brian David Josephson postulated in his master's thesis in 1962, that tunneling of Cooper pairs between two superconductors is possible without dissipation of energy. His publication in the same year [1] motivated an entire field of research and will still be of great interest for many years to come. He predicted that the tunneling of Ψ_1 and Ψ_2 would result in a superconducting current

$$j = j_\mathrm{c} \cdot \sin(\varphi_{\mathrm{sc},2} - \varphi_{\mathrm{sc},1}) = \frac{I_\mathrm{c}}{A} \cdot \sin\varphi, \tag{2.8}$$

flowing across a potential barrier, where $\varphi_{\mathrm{sc},1}$ and $\varphi_{\mathrm{sc},2}$ denote the phases of the wavefunctions Ψ_1 and Ψ_2 to each side of the barrier, j_c is the maximum current density that can be carried without a voltage drop and A the junction area. Besides this *first Josephson equation*, he also predicted that the phase difference $\varphi = \varphi_{\mathrm{sc},2} - \varphi_{\mathrm{sc},1}$, for two equal superconductors and no external magnetic field, would change over time in proportion to the voltage V across the junction:

$$\frac{\partial \varphi}{\partial t} = \frac{2e}{\hbar}V = \frac{2\pi}{\Phi_0}V. \tag{2.9}$$

In the case of an external magnetic field unequal to zero, the gauge-invariant phase difference is given by:

$$\varphi = \varphi_{\mathrm{sc},2} - \varphi_{\mathrm{sc},1} - \frac{2e}{\hbar} \cdot \int_1^2 \vec{A}\,\mathrm{d}\vec{l}, \tag{2.10}$$

where $\vec{A}$ is the magnetic vector potential and $\mathrm{d}\vec{l}$ is the line integral.

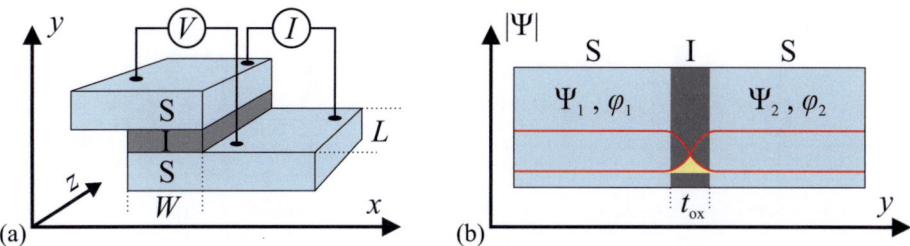

Figure 2.1.: (a) Schematic of an overlap Superconductor - Insulator - Superconductor (SIS) Josephson junction (JJ). (b) Schematic cross-section of a SIS JJ along the direction of current flow across the junction. The overlapping macroscopic wavefunctions $\Psi_1 = \sqrt{n_{s,1}}\exp(i\varphi_{sc,1})$ and $\Psi_2 = \sqrt{n_{s,2}}\exp(i\varphi_{sc,2})$ are indicated in red (solid lines), the green shading represents the coupling of the wavefunctions.

Integration of Eq. (2.9) and substitution in Eq. (2.8) leads to

$$I_s = I_c \cdot \sin\left(\frac{2\pi}{\Phi_0}V \cdot t + \varphi_0\right), \qquad (2.11)$$

where t denotes time. Accordingly, a voltage across a Josephson junction (JJ) results in an oscillating current of the frequency

$$f_J = \frac{1}{\Phi_0}V, \qquad (2.12)$$

which is known as the Josephson frequency. When applying 1 mV to a JJ, photons are emitted at a frequency of $f_{J,1\,\mathrm{mV}} \approx 483.5979\,\mathrm{GHz}$. Inversely, incident photons cause constant voltage steps across JJs, so-called *Shapiro steps*, the magnitude of which is solely dependent on the incident microwave frequency, h and the elementary charge e [2]. This *inverse Josephson effect* has been used since 1990 for the definition of the unit *volt*, by applying an rf signal to a large number of JJs connected in series [3–6].

Already in 1963, Anderson and Rowell succeeded in demonstrating the Josephson effect [32], using junctions based on a multilayer of lead and tin, separated by a thin oxide layer. Today, Josephson junctions are most commonly fabricated from multilayers of niobium and aluminum oxide. The fabrication processes relevant to and developed within this work will be discussed precisely in chapter 3 for different materials and junction shapes. Fig. 2.1 (a) depicts a 3D schematic of an overlap junction, whereas Fig. 2.1 (b) depicts the cross-section along the direction of the current flow. The macroscopic wavefunctions $\Psi_{1,2}$ in the superconducting electrodes $SC_{1,2}$ (labeled S) are depicted as solid red lines, overlapping in the region of the tunneling barrier with thickness t_{ox} (labeled I). The coupling of Ψ_1 and Ψ_2 is indicated schematically with green shading. Due to the decay of the wavefunctions within the barrier, the thickness of this barrier needs to sufficiently thin $t_{ox} < \xi$ in order to ensure coupled superconducting systems.

The material choice of the barrier is not limited to insulators but may also be normal met-

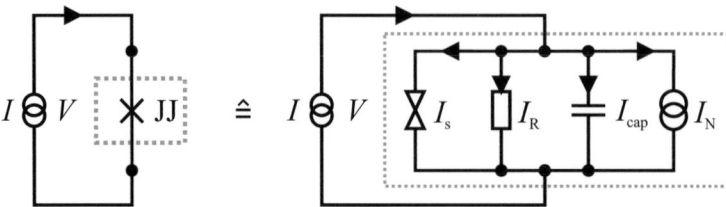

Figure 2.2.: In circuit schematics, JJs are commonly represented as a single cross as shown on the left. The corresponding resistively and capacitively shunted junction (RCSJ) Model of a single Josephson junction is shown on the right.

als (SNS junctions), typically resulting in bijective current voltage characteristics (IVCs), or ferromagnets. SFS junctions, where the barrier material is a ferromagnetic material, recently emerged as a candidate for fabrication of memory cells in superconducting electronics [33]. Within the ferromagnetic barrier the sign of the order parameter Ψ depends on the barrier thickness. This allows alteration of the JJ characteristics from conventional 0-JJs, described by Eq. (2.8) to a so-called π-JJ, where the phase undergoes a π phase shift. Such π-JJs can be described by $I_s = I_c \sin{(\varphi + \pi)} = -I_c \sin{\varphi}$ [34]. However, all trilayers dealt with in this thesis, are Superconductor - Insulator - Superconductor (SIS) multilayers, unless otherwise indicated.

The resistively and capacitively shunted junction (RCSJ) Model [35–37], developed in 1968, is most commonly used to describe the dynamic behavior of Josephson junctions. In the following discussion of the dynamics of short JJs, the Josephson phase φ is assumed to be constant over the entire junction area, before the special case of *long* Josephson junctions is discussed in section 2.3. Besides the supercurrent I_s described in Eq. (2.11), the model considers a resistive current I_R, accounting for the dissipative current transport via quasi-particles, a displacement current I_{cap} and a noise current I_N. Fig. 2.2 shows the equivalent circuit diagram of the RCSJ model. At this point it should be mentioned, that the quasiparticle tunneling shows complex dependence on the voltage across the junction. However, this aspect will be addressed later, while for the derivation of the RCSJ model, the resistance R will be simplified to be ohmic. Therefore, using Kirchhoff's law and the second Josephson equation in the form $V = \frac{\Phi_0}{2\pi} \cdot \dot{\varphi}$, the total current I through the JJ can be expressed by

$$
\begin{aligned}
I &= I_s + I_R + I_{cap} + I_N \\
&= I_c \cdot \sin{\varphi} + \frac{V}{R} + C\dot{V} + I_N \\
&= I_c \cdot \sin{\varphi} + \frac{\Phi_0}{2\pi R} \cdot \dot{\varphi} + \frac{C\Phi_0}{2\pi} \cdot \ddot{\varphi} + I_N,
\end{aligned}
\tag{2.13}
$$

where C is the inherent junction capacitance due to its parallel plate capacitor geometry as depicted in Fig. 2.1. $\dot{\varphi}$ and $\ddot{\varphi}$ denote the first and second time derivatives of the phase difference across the junction. Taking the current in units of I_c ($\gamma = I/I_c$, $\gamma_N = I_N/I_c$), voltages

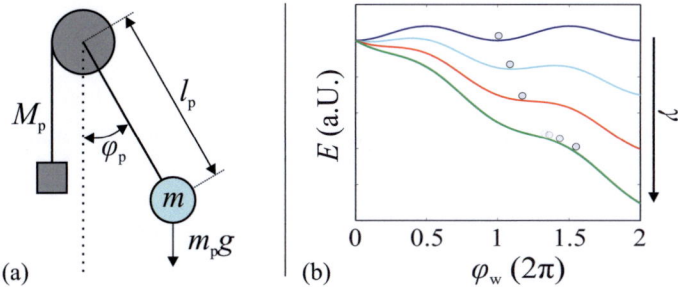

Figure 2.3.: Analogies to the dynamics of short Josephson junctions form classical physics. (a) depicts the classical pendulum, where the Josephson phase φ correspond to the angle of twist φ_p and (b) depicts the particle in a tilted washboard potential, where φ corresponds to the coordinate φ_w along the abscissa.

in units of $V_c = I_c R$ and normalizing the time to $\tau_c = \Phi_0/2\pi I_c R$, reduces Eq. (2.13) to the unit-less equation

$$\gamma = \sin\varphi + \dot{\varphi} + \beta_c \ddot{\varphi} + \gamma_N. \tag{2.14}$$

Here β_c is the so-called Stewart-McCumber parameter

$$\beta_c = \frac{2\pi I_c R^2 C}{\Phi_0}, \tag{2.15}$$

describing the damping of the junction. For values of $\beta_c > 1$ the current-voltage characteristic of the JJ is hysteretic; the junction is *underdamped*. $\beta_c \leq 1$ results in a bijective characteristic; the junction is *overdamped*. Depending on the application of the JJ, a precise tuning of β_c, to the under- or overdamped regime, is necessary. For example, most superconducting quantum interference devices (SQUIDs) include overdamped junctions, while harmonic SIS mixer elements are typically strongly underdamped devices. The tuning from $\beta_c \gg 1$ to $\beta_c \leq 1$ may be done by varying the junction capacitance C or resistance R. Using external circuitry, C can only be enlarged by a shunt capacitance C_{shunt}. Furthermore, for a given critical-current density j_c, the variation of C during the junction design directly affects the critical current I_c. Therefore the typical tuning parameter chosen for the variation of β_c is R. A shunt resistor R_{shunt}, connected in parallel to the junction, reduces the total resistance $R_{tot}^{-1} = R^{-1} + R_{shunt}^{-1}$ and hence allows precise tuning of β_c.

Analog systems from classical physics are the pendulum and the particle in a tilted washboard potential as depicted in Fig. 2.3 (a) and (b). Both analogies have proven to be very valuable for the phenomenological understanding of the Josephson junction behavior. The pendulum can be described by

$$M_p = m_p g l_p \cdot \sin\varphi_p + \Gamma_p \dot{\varphi}_p + \Theta_p \ddot{\varphi}_p, \tag{2.16}$$

Table 2.1.: Corresponding terms of Eq. (2.13) to the pendulum Eq. (2.16) and tilted washboard potential Eq. (2.17) analogies.

Josephson junction	Pendulum	Tilted washboard
phase φ	angle φ_p	coordinate φ_w
phase change $\dot{\varphi} = \frac{2\pi}{\Phi_0} V$	angular speed $\dot{\varphi}_p$	speed $v = \dot{\varphi}_w$
current I	driving torque M_p	driving force $-\frac{\partial E(\varphi_w)}{\partial \varphi_w}$
conductance $\frac{\Phi_0}{2\pi} \frac{1}{R}$	damping Γ_p	friction η
capacitance $\frac{\Phi_0}{2\pi} C$	torque of inertia Θ_p	mass m_w

where M_p is the external driving torque and the righting moment is the length of the pendulum l_p times the mass m_p, the gravitation constant g and the sine of the angle of twist φ_p. Γ_p and Θ_p are the damping constant and the torque of inertia respectively. The external driving torque M_p corresponds to the current I applied to the JJ and the angle of twist φ_p to the Josephson phase φ. Once the driving force is strong enough to generate a full turn of the pendulum, the twist angle is unequal zero, which corresponds to a voltage drop across the JJ. Depending on the torque of inertia Θ_p and the damping Γ_p, the pendulum may stay in revolving motion once having experienced a full turn even for small external driving forces. This behavior corresponds to a hysteresis in a junction's *IVC*. For large torques of inertia and low damping, *i.e.* large junction capacitance and resistance, the *IVC* exhibits a strong hysteresis, whereas it becomes non-hysteretic ($\beta_c \leq 1$) for a small capacitance and resistance.

The particle in the tilted washboard potential can be described by:

$$0 = \frac{\partial E(\varphi_w)}{\partial \varphi_w} + \eta \dot{\varphi}_w + m_w \ddot{\varphi}_w. \tag{2.17}$$

Here the particle has the mass m_w, is exposed to the friction coefficient η and moves along the coordinate φ_w in a potential described by:

$$E(\varphi_w) = E_J \left(1 - \cos \varphi_w - \gamma \varphi_w \right). \tag{2.18}$$

$E_J = \Phi_0 I_c / 2\pi$ is called the Josephson coupling energy and is a measure for the coupling strength of the wavefunctions Ψ_1 and Ψ_2. In the tilted washboard potential, the Josephson phase is mapped to a phase particle. Trapped in a potential well, it stays still, *i.e.* $\dot{\varphi} = 0$, meaning that the junction remains in the zero-voltage state. The tilt angle at which the particle is set into motion corresponds to the critical current. Whether or not the particle continues rolling down the washboard even for smaller angles, depends on its mass m_w and the friction η it experiences, *i.e.* the capacitance and conductance of the JJ. Apparently all

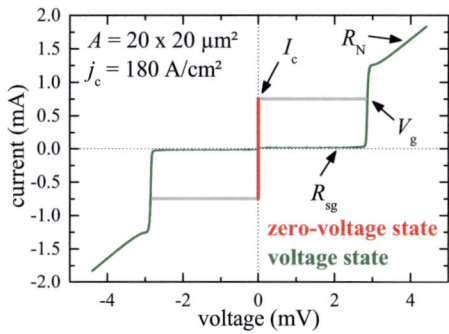

Figure 2.4.: Strongly underdamped IVC of a Josephson junction with an area of 20 x 20 μm² and $j_c = 180\,A/cm^2$. Indicated are the characteristic values, extracted for determining the quality of an underdamped JJ.

the coefficients from Eq. (2.16) and Eq. (2.17) can be mapped to Eq. (2.13) as indicated in Tab. 2.1. Consequently, the dynamics of Josephson junctions may described in the picture of a classic pendulum or a particle in a tilted washboard potential.

Fig. 2.4 shows a typical IVC of a strongly underdamped Josephson junction ($\beta_c \gg 1$), where characteristic measures are indicated. The critical current I_c is taken at the onset where $V \geq 0\,V$ and from I_c the critical-current density $j_c = \frac{I_c}{A}$ may be calculated, where A is the effective area of the junction. Just above I_c the voltage jumps from zero to the gap voltage

$$V_g = \frac{2\Delta(T)}{e}, \tag{2.19}$$

where $\Delta(T)$ is the superconducting energy gap at a given temperature T. V_g is a quality-measure for the superconducting electrodes, since it is directly dependent on the supercon-ducting energy gap in the vicinity of the tunneling barrier. It is typically above 2.7 mV for niobium electrodes of good quality. When the current through the junction is further in-creased, past the steep current-rise above the gap voltage, the IVC enters the ohmic regime, where the slope is defined by the inverse of the normal state resistance $R_N^{-1} = \left(\frac{dV}{dI}\right)^{-1}$, which is typically taken at $V = 4$ mV. From I_c and R_N the Ambegaokar-Baratoff parameter

$$I_c R_N = \frac{\pi}{2e}\Delta(T)\cdot\tanh\left(\frac{\Delta(T)}{2k_B T}\right) \tag{2.20}$$

can be calculated, which is often used to characterize the tunneling properties [38, 39]. For high-quality JJs, $I_c R_N \geq 1.5$ mV is usually achieved. For ideal Nb/AlO$_x$-interfaces, $I_c R_N$ should be 2.1 mV at $T = 4.2$ K [40]. However, in reality the interfaces will always exhibit some imperfections simply because the tunneling barrier is grown onto an aluminum wetting layer by thermal oxidation [41–43]. The remaining aluminum layer prevents the potential from being perfectly step-like in y-direction, resulting in a reduced Ambegaokar-Baratoff parameter.

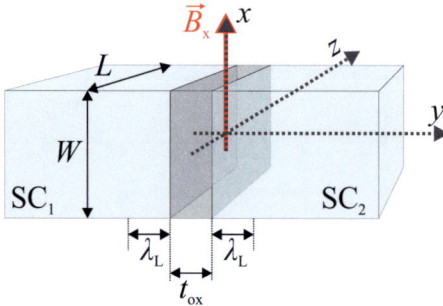

Figure 2.5.: Schematic representation of a Josephson junction penetrated by an external magnetic field in x-direction. The field decays within the SC electrodes to both sides of the tunneling barrier on a length scale of λ_L, resulting in a magnetic barrier thickness of $t_{ox,m} = t_{ox} + 2\lambda_L$ for thick electrodes of equal material.

When lowering I again, the IVC shows a hysteretic behavior for JJs with $\beta_c > 1$, *i.e.* the voltage does not drop to zero at values of $I \leq I_c$. Depending on β_c, the current at which the junction returns to the zero-voltage state, the return current I_r, may vary, decreasing for increasing β_c. For strongly underdamped JJs, the Stewart-McCumber parameter can be approximated from the measured I_c and I_r, using $\beta_c \approx \left(\frac{4}{\pi} \cdot \frac{I_c}{I_r} \right)^2$. The minimum value of I_r is temperature dependent due to thermal excitations of quasiparticles. Nevertheless, the sub-gap resistance R_{sg} divided by R_N is a common measure for the junction quality. For Nb/Al-AlO$_x$/Nb JJs, R_{sg} is measured at $2\,\text{mV}$ on the re-trapping branch and values greater than 10 indicate a good quality of the tunneling barrier. One major drawback of this convention is the dependence on the used electrode materials. For high resistive electrode materials, R_N increases, lowering R_{sg}/R_N as compared to the devices of identical barrier material and lateral dimensions but low resistive electrode materials. In order to allow a reasonable comparison of JJs of different electrode materials

$$V_m = I_c R_{sg} \tag{2.21}$$

is often used, since it is independent of material choice.

As a last quality measure to be introduced in the framework of this thesis, the behavior of JJs in an externally applied magnetic flux will be discussed, namely the $I_c(\Phi)$ dependence for short JJs. Let us consider a junction geometry as shown in Fig. 2.5, where an externally applied magnetic field penetrates the tunneling barrier of thickness t_{ox} and decays within the superconducting electrodes on a length scale of λ_L. The cross-section penetrated by the magnetic field is thus not confined to the thickness of the barrier t_{ox} but instead to a so-called *magnetic barrier thickness*, which is generally given by

$$t_{ox,m} = t_{ox} + \lambda_{L,1} \coth\left(\frac{d_1}{\lambda_{L,1}} \right) + \lambda_{L,2} \coth\left(\frac{d_2}{\lambda_{L,2}} \right), \tag{2.22}$$

where the sub-scripts 1 and 2 stand for the two superconducting electrodes SC_1 and SC_2 [44, 45]. In case of thick electrodes $d_{1,2} > \lambda_{L,1,2}$, Eq. (2.22) may be simplified to $t_{ox,m} \approx t_{ox} + 2\lambda_L$. Integrating over an area ranging from far inside the left electrode to far inside the right electrode and an infinitely small distance along the z axis dz, leads to an enclosed flux of $\Phi = \vec{B}_x t_{ox,m} dz$. From this we obtain

$$\frac{\partial \varphi}{\partial z} = \frac{2\pi}{\Phi_0} \vec{B}_x t_{ox,m},$$

(2.23)

which shows that the phase along z depends on the magnetic field $\vec{B}_x$. Integration of Eq. (2.23) and substitution in Eq. (2.8) results in

$$j_s(z) = j_c \sin\left(\frac{2\pi}{\Phi_0} t_{ox,m} \vec{B}_x z + \varphi_0\right).$$

(2.24)

Consequently the supercurrent density depends on the magnetic field parallel to the plane of the tunneling barrier.

In order to derive the dependence of the critical current on the magnetic flux through the junction, integration of $j_s(z)$ along the direction of the magnetic field, *i.e.* $\iota_c(z) = \int_{-W/2}^{+W/2} j_s(z) dx$, and along the z axis is necessary, and results in

$$I_s(\vec{B}_x) = \int_{-L/2}^{+L/2} \iota_c(z) \sin\left(\frac{2\pi}{\Phi_0} t_{ox,m} \vec{B}_x z + \varphi_0\right) dz$$

$$= \mathscr{F}\left\{\exp(i\varphi_0) \int_{-\infty}^{+\infty} \iota_c(z) \exp\left(i\frac{2\pi}{\Phi_0} t_{ox,m} \vec{B}_x z\right) dz\right\}.$$

(2.25)

According to Eq. (2.25) the modulation of the critical current of a short Josephson junction corresponds to the Fourier transform of the current, when penetrated by a magnetic field. In case of a spatially homogeneous critical-current density $j_c(x,z)$ in a rectangular JJ, $\iota_c(z)$ is constant and unequal zero from $-L/2$ to $+L/2$ and zero everywhere else, as depicted in the inset of Fig. 2.6. The Fourier transformation of such a function results in

$$I_c(\Phi) = I_c(0) \left|\frac{\sin\left(\pi\frac{\Phi}{\Phi_0}\right)}{\pi\frac{\Phi}{\Phi_0}}\right|,$$

(2.26)

and is well known from optics. Just like the diffraction pattern of light passing through a single slit, the $I_c(\Phi)$ modulation given in Eq. (2.26) is called the *Fraunhofer pattern*, and is shown in Fig. 2.6. The minima occur at integer multiples of the magnetic flux quantum $n\Phi_0$, where $n \in \mathbb{Z}$. Phenomenologically, these minima correspond to the points where the Josephson phase is modulated by $2\pi n$, causing cancellation of the effective superconducting

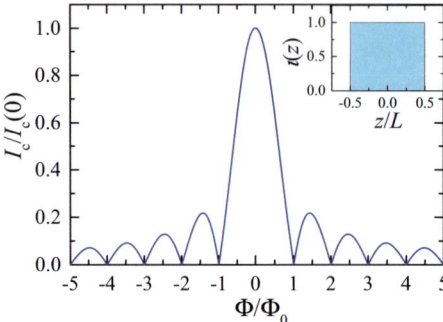

Figure 2.6.: Theoretical $I_c(\Phi)$ dependence for a short rectangular Josephson junction. The inset shows the current distribution $\iota(z)$ in a junction with an ideal tunneling barrier.

charge transport. Measurements of $I_c(\Phi)$ therefore contain information on the current distribution in the junction. For a given shape of a JJ, comparison of measurement and theory is therefore a convenient way to probe the tunneling barrier quality. A short JJ of high quality exhibits in a symmetric $I_c(\Phi)$ dependence with sharp distinct minima at $I_c(2\pi n) = 0\,\mathrm{mA}$ in the case of a well-aligned external magnetic field.

The different activation mechanisms to the voltage state of junctions shall be introduced by means of the previously presented tilted washboard potential, before the basics of long Josephson junctions are discussed in section 2.3. In the conventional *IVC*, the measured I_c corresponds to the switching current $I_{c,sw}$ at which the washboard has been tilted sufficiently, such that the resulting potential barrier U to the right of the phase particle in Fig. 2.7 (a) is smaller than the thermal energy k_BT [19, 46–48]. For typical measurements at $T = 4.2\,\mathrm{K}$, the measured critical current is therefore always smaller than the real $I_{c,0}$ at $T = 0\,\mathrm{K}$, due to thermal activation (TA) of the Josephson phase. The switching event is a stochastic process, described by the escape rate $\Gamma = \alpha_t \frac{\omega_{pl}}{2\pi} \exp\left(-\frac{U}{k_BT}\right)$. Here, $\alpha_t = \sqrt{1 + (1/2Q)^2} - 1/2Q$ is a pre-factor specific for the different damping limits [46], with $Q = \omega_{pl}RC$. This leads to switching histograms with standard deviations [49–51]

$$\sigma \propto I_c^{1/3}. \tag{2.27}$$

For most experiments the switching current $I_{c,sw}$, extracted form an *IVC*, is of sufficient accuracy and is therefore used as I_c throughout literature. Unless otherwise indicated this also applies to this work.

Fig. 2.7 (b) shows the case of resonant activation (RA). The resonant frequency ω_{pl} with which the particle may be lifted over the potential barrier, depends on the shape of the potential given in Eq. (2.18), which is in turn dependent on the applied bias current I. This frequency is equivalent to the eigenfrequency of the JJ, determined by its capacitance and

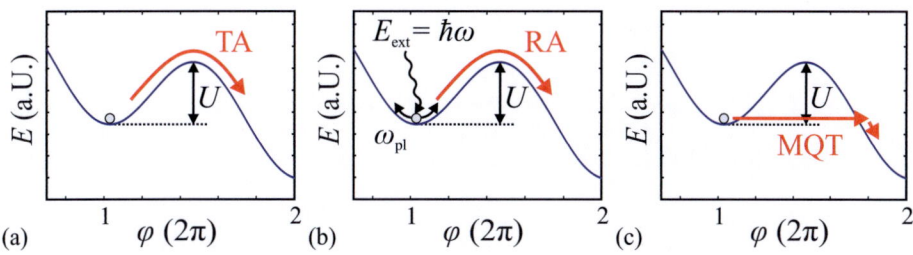

Figure 2.7.: Three different types of activation mechanisms represented in the tilted washboard potential at a bias current $I < I_{c,0}$. (a) depicts the case where the phase particle is thermally excited over the remaining potential barrier $U \leq k_B T$, i.e. thermal activation (TA); (b) the phase particle is resonantly excited to its plasma frequency ω_{pl} by an external microwave, i.e. resonant activation (RA); (c) the phase particle escapes the potential well due to macroscopic quantum tunneling (MQT).

its Josephson inductance

$$L_J = V \left(\frac{\partial I}{\partial t}\right)^{-1} = \frac{\Phi_0}{2\pi I_c \cos\varphi} = \frac{L_{J0}}{\sqrt{1-\gamma^2}}, \qquad (2.28)$$

where $L_{J0} = \Phi_0/2\pi I_c$. The plasma frequency can then be derived to

$$\omega_{pl} = \frac{1}{\sqrt{L_J C}} = \sqrt{\frac{2\pi I_c}{\Phi_0 C}} \left(1-\gamma^2\right)^{1/4} = \omega_{pl,0} \left(1-\gamma^2\right)^{1/4}, \qquad (2.29)$$

with $\omega_{pl,0} = \sqrt{2\pi I_c/\Phi_0 C}$ and $\gamma = I/I_c$ [19, 27, 47, 52]. Essentially this creates a scenario comparable to the above mentioned thermal activation, however with $E_{ext} = \hbar\omega \geq U$ supplied by an external electro-magnetic wave.

In the case of very low temperatures, typically in the mK-range, the remaining thermal energy $k_B T \leq U$ is insufficient to cause switching events. Nonetheless, switching of the JJ to the voltage state may still occur at finite values of the potential barrier U due to quantum fluctuations or tunneling [47, 51, 53–55]. The tunneling effect, schematically shown in Fig. 2.7 (c), is often called macroscopic quantum tunneling (MQT), since the phase particle represents a macroscopic measure, tunneling out from the potential well and switching the JJ to the voltage state. MQT can be observed below the *crossover temperature*

$$T^* = \alpha_t \frac{\hbar\omega_{pl}}{2\pi k_B}, \qquad (2.30)$$

where the switching is no longer temperature dependent. In the quantum regime the standard deviation and the position of the switching histogram saturate as shown in Fig. 2.8. For conclusive evidence for operating in the quantum regime, elaborate measurements are necessary, combining low-temperature measurements with microwave irradiation [47, 55]. Nonetheless, the saturation of the mean switching current I_m and the standard deviation of

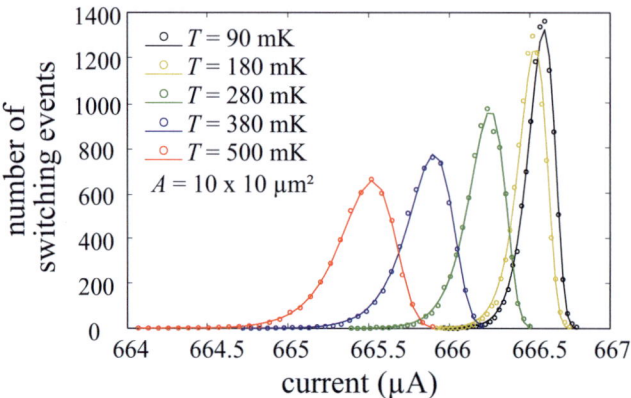

Figure 2.8.: Histograms of 10000 switching events, measured at different temperatures. For $T \lesssim 150\,\text{mK}$ the width and mean switching-current I_m of the individual histograms saturate, indicating that the activation mechanism is dominated by quantum tunneling instead of thermal excitation.

the histograms, are first indicators for having reached the quantum regime. In the example shown in Fig. 2.8, the current through $10 \times 10\,\mu\text{m}^2$ JJ has been ramped 10000 times for different temperatures from 500 mK down to 90 mK.

Conclusion

In this section, Josephson junctions have been introduced by means of the first and second Josephson equation, as well as the RCSJ Model. The dynamic behavior has been explained with the help of analogies from classical physics. Means of extracting quality parameters of a JJ from dc measurements were introduced, before the different activation mechanisms were discussed at the end.

2.3. Long Josephson Junctions

All Josephson junctions discussed in section 2.2 have been considered to be short, *i.e.* the width W and length L are such that the Josephson phase φ could be assumed to be constant across the entire JJ for no externally applied magnetic field. In this section, the length of the junctions will be extended, while the width remains short, resulting in a one dimensional system, where the phase may vary along the coordinate z. Such a junction is called a *long* Josephson junction (LJJ) and requires the length to exceed the so-called Josephson penetration depth

$$\lambda_\text{J} = \sqrt{\frac{\Phi_0}{2\pi\mu_0 t_{\text{ox,m}} j_\text{c}}}. \tag{2.31}$$

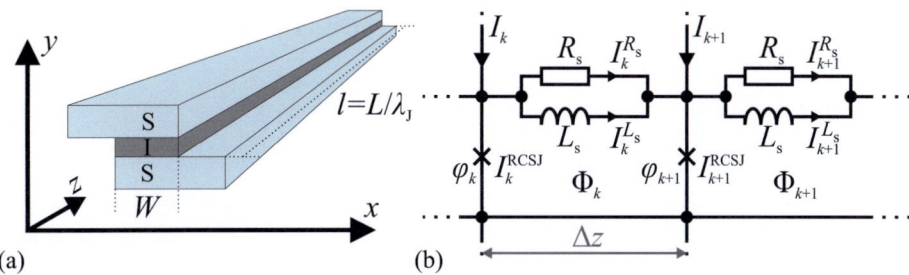

Figure 2.9.: (a) Schematic representation of a LJJ. (b) Equivalent circuit diagram of a LJJ. Each short interlinked JJ$_k$ is described by the RCSJ model introduced in Eq. (2.13)

For such LJJs the eigenfields created by the bias current have to be taken into account. This will be examined further when flux-flow steps are introduced at the end of this section. Fig. 2.9 (a) shows a schematic of a long JJ, together with the equivalent circuit diagram typically used for modeling long junctions in Fig. 2.9 (b) [56]. For simplicity, the bias current distribution is assumed to be constant, which is in fact not the case for most real devices. The circuit consists of numerous short JJs, Δz apart from each other and interlinked by a surface resistance R_s and inductance L_s connected in parallel. Each of these short JJs, can be described by the RCSJ model introduced in section 2.2. The bias current fed to each short junction (JJ$_k$) is given by I_k, where $k \in \mathbb{N}$. Due to flux quantization, the phase change from JJ$_k$ to JJ$_{k+1}$ can be described by

$$\varphi_{k+1} - \varphi_k = \frac{2\pi}{\Phi_0} \left(\Phi_{\text{ext}} - L_s I_k^{L_s} \right), \tag{2.32}$$

where the external flux $\Phi_{\text{ext}} = t_{\text{ox,m}} \vec{B}_x \Delta z$ may be due to external fields as well as eigenfields of the bias current. Using Kirchhoff's law at the point where I_{k+1} is fed to the LJJ, we get

$$I_k^{R_s} + I_k^{L_s} + I_{k+1} = I_{k+1}^{R_s} + I_{k+1}^{L_s} + I_{k+1}^{\text{RCSJ}}. \tag{2.33}$$

The combination of Eq. (2.32) and Eq. (2.33), expressed in the differential limit $\Delta z \to 0$, results in

$$\frac{\varphi_{k+1} - \varphi_k}{\Delta z} \overset{\Delta z \to 0}{=} \frac{\partial \varphi}{\partial z} = \frac{2\pi}{\Phi_0} \left(t_{\text{ox,m}} \vec{B}_x - L^\star I_k^{L_s} \right), \tag{2.34}$$

and

$$\frac{I_{k+1}^{L_s} - I_k^{L_s}}{\Delta z} \overset{\Delta z \to 0}{=} \frac{\partial I^{L_s}}{\partial z} = \left(j^\star - j^{\star,\text{RCSJ}} \right) W - \frac{\partial I^{R_s}}{\partial z}. \tag{2.35}$$

The specific inductance $L_s/\Delta z$ is given by $L^\star = \mu_0 t_{\text{ox,m}}/W$, whereas $j^\star$ and $j^{\star,\text{RCSJ}}$ denote the bias current density $I_k/(W\Delta z)$ and the current density $I_k^{\star,\text{RCSJ}}/(W\Delta z)$ for each short JJ$_k$, respectively. A further derivation of Eq. (2.34) with respect to z results in:

$$\frac{\partial^2 \varphi}{\partial z^2} = \frac{2\pi}{\Phi_0}\left(t_{\text{ox,m}}\frac{\partial \vec{B}_x}{\partial z} - L^\star \frac{\partial I^{L_s}}{\partial z}\right). \tag{2.36}$$

When solving Eq. (2.36) for $\partial I^{L_s}/\partial z$ with a homogeneous field applied, it can be substituted in equation Eq. (2.35). Additionally $I^{R_s} = -1/R_s \cdot (\partial V/\partial z) = -W/\zeta_s \cdot (\partial V/\partial z)$ can be substituted into Eq. (2.13) expressed in terms of the RCSJ current density $j^{\star,\text{RCSJ}} = I/WL$ [56]. This allows derivation of:

$$\frac{\Phi_0}{2\pi L^\star}\frac{\partial^2 \varphi}{\partial z^2} = j_c \sin\varphi - j + \frac{V}{\zeta} + C^\star \frac{\partial V}{\partial t} - \frac{1}{\zeta_s}\frac{\partial^2 V}{\partial z^2}. \tag{2.37}$$

Using the second Josephson equation Eq. (2.9) to express the voltages in terms of the Josephson phase, one gets the perturbed Sine-Gordon equation (pSGE):

$$\frac{\Phi_0}{2\pi L^\star}\frac{\partial^2 \varphi}{\partial z^2} - \frac{\Phi_0 C^\star}{2\pi}\frac{\partial^2 \varphi}{\partial t^2} - j_c \sin\varphi = -j + \frac{\Phi_0}{2\pi\zeta}\frac{\partial \varphi}{\partial t} - \frac{\Phi_0}{2\pi\zeta_s}\frac{\partial^3 \varphi}{\partial z^2 \partial t}, \tag{2.38}$$

where $\zeta = R \cdot W\Delta z$ is the sheet resistance and $C^\star$ is the specific capacitance C/WL. Just like in equation Eq. (2.13), here the time can also be normalized to the inverse plasma frequency ω_{pl}^{-1} and the coordinate z to the Josephson penetration depth λ_J. Consequently, Eq. (2.38) can be reduced to

$$\varphi_{\tilde{z}\tilde{z}} - \varphi_{\tilde{t}\tilde{t}} - \sin\varphi = -\gamma + \alpha\varphi_{\tilde{t}} - \beta\varphi_{\tilde{t}\tilde{z}\tilde{z}}, \tag{2.39}$$

where the sub-scripts $\tilde{z}$ and $\tilde{t}$ stand for derivatives with respect to the coordinate $\tilde{z} = z/\lambda_J$ and time $\tilde{t} = \omega_{\text{pl}}t$, respectively. $\alpha = 1/\sqrt{\beta_c}$ and β are dimensionless damping parameters. For regular Nb/Al-AlO$_x$/Nb technology, $\beta = \omega_{\text{pl}}L^\star/\zeta_s$, accounting for the surface losses, can be neglected, since they are very low as compared to α [56]. Thus, Eq. (2.39) may be further simplified to a second order differential equation in most cases.

Based on the equivalent circuit diagram depicted in Fig. 2.9 (b), the pendulum, as well as the washboard analogy, introduced in section 2.2, can be extended by the coordinate z. In the case of the pendulum analogy, this results in a chain of pendula, linked by torsional springs and each short JJ$_k$ is described by Eq. (2.14). The linking springs represent the inductances L_s. For the washboard analogy, the identical extension along the z axis applies. The phase is then represented by a chain of phase particles connected to each other by a flexible band. In both cases, the correspondences from Tab. 2.1 apply, however with a phase being dependent on z. It should be noted that the accuracy of these analogies, obviously depends on the discretization of the model. The characteristic length scale on which the extended phase may change is given by the Josephson penetration depth λ_J introduced in Eq. (2.31).

Due to the z dependence of the phase, there exist more possibilities of excitations in long

JJs than in short JJs. In the following, the excitations relevant to this work will be introduced and discussed. Eq. (2.39) is clearly a non-linear wavefunction, well-known from classical physics and mathematics. The first type of excitation is given for small amplitudes $\mathscr{A}$ and phase differences $\varphi \approx \sin\varphi$ in strongly underdamped ($\beta_c \gg 1$) junctions of infinite length, Eq. (2.37) simplifies to the linearized unperturbed SGE

$$\lambda_J^2 \varphi_{zz} - \omega_{pl}^{-2} \varphi_{tt} - \varphi = 0. \tag{2.40}$$

The solution of which is given by:

$$\varphi(z,t) = \mathscr{A} \cdot \exp\left[i\left(\vec{k}z - \omega t\right)\right] + \arcsin\gamma. \tag{2.41}$$

The dispersion relation

$$\omega(\vec{k},\gamma) = \left[c_{sw}^2 \vec{k}^2 + \omega_{pl,0}^2 \cdot \sqrt{1-\gamma^2}\right]^{-1/2} \tag{2.42}$$

can be derived by substituting Eq. (2.41) in Eq. (2.40), with $\vec{k}$ being the wave vector and $c_{sw} = \omega_{pl}\lambda_J$ the *Swihart velocity* [57]. Essentially, c_{sw} is the speed of light in a medium, *i.e.* the maximum velocity an electro-magnetic wave can reach inside a LJJ. In fact, this velocity is at most a few percent of the speed of light in vacuum. According to Eq. (2.42) only waves of frequency $f \geq \omega_{pl}/2\pi$ can propagate along the LJJ, while waves of lower frequency are damped exponentially.

Even stronger restrictions apply for JJs of finite length $l = L/\lambda_J$, where the boundary conditions $\varphi_{\bar{z}}(0) = \varphi_{\bar{z}}(l)$ for linear junctions and $\varphi(0) = \varphi(l)$ and $\varphi_{\bar{z}}(0) = \varphi_{\bar{z}}(l)$ for annular junctions lead to discrete values of $\vec{k}$

$$\vec{k}_n = \begin{cases} (\pi/L)\,n, & \text{for linear LJJs} \\ (\pi/L)\,2n, & \text{for annular LJJs,} \end{cases} \tag{2.43}$$

causing standing waves in the LJJs.

The second type of excitation is eponymous to this thesis, describes a full twist of φ along z by 2π and is called a *Fluxon* or *Josephson vortex*. Let us consider the chain of pendula, fixed at one end and experiencing a full $360°$ twist at the other. When following the path of the individual pendula, they describe a spiral. For a long chain, this twist can be moved along z, without the shape changing, nor the amplitude decreasing. The twist behaves just like a particle and may only escape the system at the ends of linear systems, or not at all from annular systems [19]. This type of particle is also known as a soliton from mathematics and classical physics [58]. Due to their unique and interesting characteristics, such fluxons have been studied extensively over the last decades [19, 27, 57, 59–62].

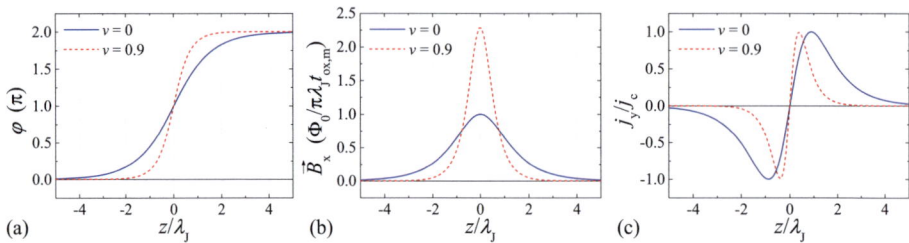

Figure 2.10.: (a) Phase response, (b) magnetic field and (c) supercurrent of a Josephson vortex. The solid blue lines correspond to the stationary solution of a still-standing vortex, while the dashed red lines depict the characteristics of a vortex moving at a speed of 90% of the Swihart velocity.

In a long Josephson junction such fluxons are best derived by looking at the unperturbed Sine-Gordon equation (uSGE)

$$\varphi_{\tilde{z}\tilde{z}} - \varphi_{\tilde{t}\tilde{t}} + \sin\varphi = 0 \qquad (2.44)$$

with the boundary conditions for an infinitely long JJ

$$\lim_{z\to\pm\infty} \varphi(z) = n \cdot 2\pi, \text{ and}$$
$$\lim_{z\to\pm\infty} \varphi_z(z) = 0. \qquad (2.45)$$

Eq. (2.44) has the solution:

$$\varphi(\tilde{z},\tilde{t}) = 4\arctan\left[\exp\left(\pm\frac{\tilde{z} - v\tilde{t} - \tilde{z}_0}{\sqrt{1 - v^2}}\right)\right], \qquad (2.46)$$

describing the phase of a fluxon at $\tilde{z}_0(t) = v\tilde{t}$, where v is a fluxon velocity normalized to c_{sw}; $v \in (-1,1)$. The velocity of a moving fluxon depends on the damping parameter α and the applied bias current in the form of [63]

$$v = \left[1 + \left(\frac{4\alpha}{\pi\gamma}\right)^2\right]^{-1/2}. \qquad (2.47)$$

For high-quality Nb/Al-AlO$_x$/Nb JJs, α is in the range of approximately 0.1, allowing very high speeds of vortices inside a LJJ close to the Swihart velocity. The solution Eq. (2.46) of the uSGE describes a phase variation of $\pm 2\pi$ around its center and is shown in Fig. 2.10 (a) as a solid blue line. The dashed red line depicts a vortex moving at a speed of $v = 0.9$. Due to the Lorentz force, caused by the bias current, the fluxon experiences a relativistic length-contraction. From Eq. (2.23) we know that a change in $\varphi(z)$ by 2π corresponds to a magnetic field $\vec{B}_x$ carrying the flux Φ_0. Again, the solid blue line in Fig. 2.10 (b) depicts the resting case, while the dashed red line depicts the moving vortex. The integral $\int_{-\infty}^{+\infty} \vec{B}_x dz$ in both cases equals the magnetic flux quantum. In Fig. 2.10 (c) the supercurrent of a fluxon is shown for a stationary vortex (solid blue line) and a moving vortex (dashed red line). The current circulates around its center, upholding the magnetic field. Depending on its sign $+/-$, the described vortex is called a fluxon/vortex or anti-fluxon/-vortex respectively. Due to their solitonic behavior, characteristics such as mass and momentum may be attributed to Josephson vortices [59].

Some particular effects of fluxons shall be highlighted at this point. According to the second Josephson equation, a variation of φ over time corresponds to a voltage drop across the JJ. This means that a moving vortex corresponds to a localized voltage pulse traveling along the LJJ. The higher the speed of the vortex, the sharper the voltage pulse is. In an annular LJJ, with a trapped fluxon moving at a speed v along the circular path, it is therefore possible to measure the mean voltage over time $\langle V \rangle = \dot{\varphi} \left(\Phi_0/2\pi \right) = v c_{sw} \Phi_0/\mathscr{L}$, where $\mathscr{L} = 2\pi r$ is the mean circumference of the annular JJ and r is the radius. With increasing speed, *i.e.* $I \to I_c$, $\langle V \rangle$ approaches $c_{sw} \Phi_0/\mathscr{L}$ asymptotically. An increasing number of fluxons inside the LJJ multiplies the voltage. These voltages are known as *zero field steps* and may be observed in the *IVC* at

$$V_{ZFS} = \frac{c_{sw} \Phi_0}{\mathscr{L}} \cdot n. \tag{2.48}$$

In section 2.2 it has been established that the Josephson phase may be altered by applying an external magnetic field parallel to the plane of the tunneling barrier. Accordingly, the phase in a long JJ can also be altered by $\vec{B}_x$. Unlike $I_c(\Phi)$ of a short JJ, the dependence for a long JJ does not correspond to a Fraunhofer pattern. Instead, screening currents circulating in the yz-plane will keep the junction in the diamagnetic Meissner phase at first. Consequently, the measurable critical current drops linearly for low magnetic fields. At the point where the screening currents are strong enough to create fluxons, vortices will enter the JJ at the ends. In contrast to the Fraunhofer pattern shown in Fig. 2.6, the minima do not occur at $n\Phi_0$ and the current is not necessarily zero. Fig. 2.11 (a) shows a typical dependence of the critical current of a LJJ under the influence of an external magnetic field, applied by a coil magnet. The dependence has been recorded for positive and negative bias currents (depicted in black and red, respectively). The highly symmetric curve with respect to the bias-current axis, indicates that the eigenfields affecting the JJ are minimal and that there is no flux trapped inside the junction. With increasing normalized lengths l, the $I_c(\Phi)$ dependences get more narrow since the Josephson phase $\varphi(z)$ becomes more susceptible to magnetic fields.

In accordance with the scenario described above, plasma waves of the wavelength $\lambda = nL/2 = n\pi/k$ can be created in short and long JJs. These plasma waves appear for magnetic flux through the junction $\Phi = n\Phi_0/2$, which corresponds to the magnetic field, where the critical current is minimal in the $I_c(\Phi)$ dependence. There, the frequency of the Joseph-

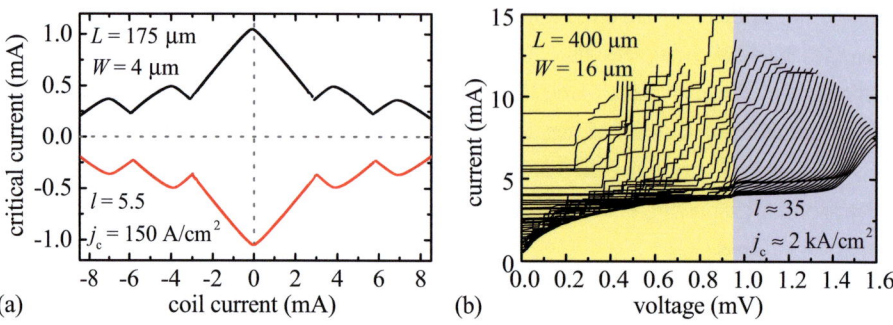

Figure 2.11.: Typical measurement of a $I_c(\Phi_0)$ dependence of a long Josephson junction is shown in (a). The normalized length of this sample has been $l = 5.5$ at a critical-current density of $j_c = 150\,\mathrm{A/cm^2}$. In (b) an *IV* family under the influence of an externally applied magnetic field is shown. The Fiske regime is highlighted in yellow and the flux-flow regime in blue.

son currents matches the one of the electro-magnetic cavity formed by the electrodes and the insulating barrier. Hence, voltage steps occur in the *IVC* at

$$V_n = \frac{\omega_{pl}}{2\pi} \frac{\Phi_0 \lambda_J}{2L} n. \tag{2.49}$$

These steps were first observed by M. D. Fiske in 1964 and are therefore called *Fiske steps* [64, 65]. Typical values for V_n are in the µV range and correspond to frequencies of a few 100 GHz. In Fig. 2.11 (b) a family of *IVC*s under the influence of an externally applied magnetic field is shown. The region where Fiske steps occur is highlighted in yellow. Measurements of Fiske steps are often used to extract the specific capacitance

$$C^\star = \frac{\pi j_c \Phi_0 \lambda_J^2}{2L^2} \left(\frac{n}{V_n}\right)^2 \tag{2.50}$$

of trilayers.

Finally, the so-called *flux-flow steps* will be introduced in this chapter. In a long JJ exposed to a magnetic field which increases slowly from zero, the JJ will remain in the Meissner phase at first. However, unlike in pure superconductors, the screening currents do not decay exponentially on a length scale of λ_L, but rather on a scale of λ_J along z, which is orders of magnitude larger than the London penetration depth. Mathematically, this can be derived by the time invariant version of Eq. (2.40) and $\varphi_{\bar{z}}(0) = \varphi_{\bar{z}}(l) = 2\pi/\Phi_0 t_{ox,m} \vec{B}_x$. An increasing magnetic field will cause fluxons to enter from the ends of the junction. If an applied bias current is strong enough to overcome the pinning force, these fluxons will move at a velocity v. This results in fluxons arriving at the end of the junction at a frequency of $f = {}^{vc_{sw}}/\delta$, where δ is the distance between two neighboring fluxons. At any fixed point z_f along the junction the mean voltage drop is therefore given by $V_{FF} = {}^{vc_{sw}} \Phi_0/\delta$, since every passing fluxon causes $\varphi(z_{\rm f})$ to increase by 2π. For an integer number of fluxons inside the JJ, each confined to a distance δ, it follows that $\Phi_0 = t_{\rm ox,m}\vec{B}_x\,(L/n)$. The voltage of flux-flow steps is therefore given by

$$V_{\rm FF} = t_{\rm ox,m}B_x v c_{\rm sw} = t_{\rm ox,m}B_x c_{\rm sw}\left[1+\left(\frac{4\alpha}{\pi\gamma}\right)^2\right]^{-1/2}. \qquad (2.51)$$

The maximum flux flow voltage is again reached as $v \to 1$ [19]. As compared to Fiske steps, flux-flow steps occur at higher voltages and typically exhibit a larger differential resistance in the *IVC*. In Fig. 2.11 (b) flux-flow steps are highlighted in blue. Both the Fiske and the flux-flow steps are used for the realization of local oscillators based on long Josephson junctions and will be further discussed in subsection 3.4.5.

Conclusion

In the field of fundamental physics, Josephson vortices are of great interest, because they allow the investigation of isolated particles in a one dimensional system, or even interaction between multiple particles [60, 61]. More recently, a new type of Josephson vortex attracted attention, the so-called fractional Fluxon or κ-Fluxon, carrying merely a fraction of the magnetic flux quantum Φ_0. This is realized by a pair of current injectors attached to one of the junction's electrodes and was first proposed in 2002 [16, 17, 66]. κ-fluxons, and the realization thereof, will be discussed further in subsection 3.4.2.

In applied physics, long Josephson junctions and vortex dynamics therein, have gained much attention since they were first proposed as local oscillators by Nagatsuma *et al.* in 1983 [8]. More precisely, the directional flow of fluxons under an applied magnetic field and bias current is used to emit electro-magnetic radiation from one end of the LJJ. The emitted frequency may be tuned by the dc currents supplying $\vec{B}_x$ and the bias current. Fabrication and possible improvements to the state of the art flux-flow oscillators will be addressed further in subsection 3.4.5.

2.4. Josephson Junction Devices

In this section the theoretical behavior of the Josephson junction devices, developed within this thesis, will be described. The first part deals with superconducting quantum interference devices (SQUIDs) which not only play a significant role for the precise determination of fabrication parameters, such as mutual inductances in superconducting striplines [67, 68], but are also extremely sensitive detector devices for small magnetic fields in the range of the magnetic flux quantum Φ_0.

In the second part the possibility of creating *fractional* Josephson vortices in long Josephson junctions by applying an additional dc current to injectors attached to one of the junctions electrodes, is described. Using such devices, vortices carrying arbitrary fractions of Φ_0 can be created artificially, altering the Josephson phase along the long dimension of the junction. When placing two of such κ-Fluxons along a long junction, double-well potentials can be created, forming very promising candidates for a new type of qubits. Furthermore, a device incorporating multiple injector-pairs has been proposed for creation of metamaterials with band structures that are tunable during experiments [69].

The last part of this section is dedicated to integrated receiver devices composed of both short and long junctions. Short junctions with lateral dimensions much below λ_J, and high critical-current densities form the mixer device for electro-magnetic waves in the sub-mm range, while long junctions may act as an on-chip local oscillator (flux-flow oscillator) for pumping of the mixers. In this thesis the main focus concerning FFOs, is on the development of LC-shunted oscillators with reduced linewidth of the emitted electro-magnetic radiation, as compared to conventional FFOs.

2.4.1. Superconducting Quantum Interference Devices

Over the last decades SQUIDs have been investigated intensively and consequently numerous realizations have been developed [13, 14, 19, 70][RNM$^+$12]. They combine the flux quantization in superconducting ring structures and the Josephson effect introduced in section 2.1. The most common SQUID is the symmetric dc SQUID based on a superconducting ring, interrupted by two Josephson junctions as shown schematically in Fig. 2.12 (a). In this case, the junctions have identical parameters, leading to $I_0 = (I_{c,1}+I_{c,2})/2$, $C_{sq} = (C_1+C_2)/2$, $L_{sq} = L_1 + L_2$ and $R_{sq} = 2R_1R_2/(R_1+R_2)$, such that the bias current I splits equally to the left and right SQUID arm. Here, C_{sq}, L_{sq} and R_{sq} are the SQUID's capacitance, inductance and twice the resistance, respectively. Each of the JJs can be described by the RCSJ model as depicted in Fig. 2.12 (b). When a magnetic field $\vec{B}_{ext}$, normal to the SQUID plane is applied, a superposed current I_{circ} is induced, resulting in $I_1 = I/2 + I_{circ}$ and $I_2 = I/2 - I_{circ}$, describing the total current through the junctions JJ1 and JJ2. Obviously, I_{circ} causes an additional magnetic flux $\Phi_{circ} = I_{circ}L_{sq}$, leaving the total flux through the SQUID to be $\Phi = \Phi_a + \Phi_{circ}$. Analogous to the derivations in [14, 19, 27] $\varphi_{JJ2} - \varphi_{JJ1} = 2\pi/\Phi$ follows, where φ_{JJ1} and φ_{JJ2} stand for the Josephson phases of the left and right junctions. The total current through the SQUID can be expressed by:

$$I = 2I_0 \cos\left(\pi\frac{\Phi_a}{\Phi_0}\right) \sin\left(\varphi_{JJ1} + \pi\frac{\Phi_a}{\Phi_0}\right), \tag{2.52}$$

under the assumption that $L_{sq} \to 0$. The maximum supercurrent carried by symmetric dc SQUIDs is therefore given by:

$$I_{s,max} = 2I_0 \left|\cos\left(\pi\frac{\Phi_a}{\Phi_0}\right)\right|. \tag{2.53}$$

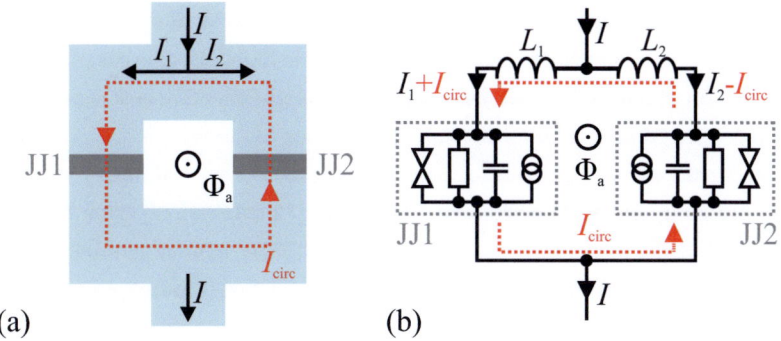

Figure 2.12.: (a) Schematic representation of a 2-JJ dc SQUID. (b) Equivalent circuit diagram of a 2-JJ dc SQUID. Each of the JJs is described by the RCSJ model introduced in Eq. (2.13).

The critical current follows a cosinusoidal dependence on the applied magnetic field. For $L_{sq} \to 0$, the supercurrent $I_{s,max}$ vanishes completely at $\Phi/\Phi_0 = (2n+1)/2$. However, in real SQUID devices the inductance L_{sq} is never zero and consequently the modulation of the critical current cannot be solved analytically anymore. Instead, it has to be solved numerically leading to the Langevin equations

$$\frac{i}{2} + i_{circ} = \sin \varphi_{JJ1} + \dot{\varphi}_{JJ1} + \beta_c \ddot{\varphi}_{JJ1} + \gamma_{N,JJ1}$$
$$\frac{i}{2} - i_{circ} = \sin \varphi_{JJ2} + \dot{\varphi}_{JJ2} + \beta_c \ddot{\varphi}_{JJ2} + \gamma_{N,JJ2},$$

(2.54)

where the currents are normalized to I_0, time to τ_c and $\varphi_{JJ2} - \varphi_{JJ1} = 2\pi \left(\Phi_a + 1/2 \cdot \beta_L i_{circ} \right)$ [13]. The screening parameter

$$\beta_L = \frac{2 I_0 L_{sq}}{\Phi_0}$$

(2.55)

describes the effect, whereby part of the applied magnetic flux will be screened from the SQUID. With an increasing SQUID inductance and critical current, this effects gets stronger, resulting in a decreasing modulation depth of $I_{s,max}(\Phi_a)$.

Depending on the damping regime of the JJs, two types of measurements can be distinguished. When using unshunted JJs, *i.e.* strongly underdamped junctions with hysteretic *IVCs*, the SQUID is typically unbiased $I = 0\,\mathrm{mA}$. To measure a magnetic field, the bias current is ramped up, and the switching current is mapped to a previously recorded $I_{s,max}(\Phi_a)$ dependence. This measurement type will be referred to as a single shot measurement within this thesis. The clear advantage is that for the most part, the SQUID is unbiased and in the zero voltage state. Consequently, the dissipated energy and heat introduced to the chip is very low. Disadvantageous is however, that after each measurement point, the bias current and in some cases also the magnetic field to be measured, needs to be set back to zero. Therefore a continuous recording of the magnetic field is impossible. For the detection of vortex states in long Josephson junctions, this single shot procedure has been employed, since minimal energy dissipation on the chip has been anticipated. It should be noted that the SQUIDs used, have been based on current-asymmetric SQUIDs, consists of 3 JJs. This results in a Φ-shift of the $I_c(\Phi_a)$ dependence, so that no additional flux bias is necessary. A more detailed description will follow in subsection 3.4.2.

The second and more common measurement type is the continuous mode, where over-damped JJs are used. As mentioned above, JJs with $\beta_c \leq 1$ exhibit a bijective IVC. When integrated in a SQUID, this allows overcritical biasing $I \geq 2I_0$, exhibiting minimal voltage drop at $\Phi_a = n\Phi_0$ for a symmetric SQUID. For $\Phi_a = (2n+1)\Phi_0/2$, the screening current I_{circ} is at its maximum, resulting in a minimized critical bias current, $i.e.$ maximal voltage drop. Hence the SQUID acts as a flux-to-voltage converter and shows a sinusoidal $V(\Phi_a)$ depen-dence. At the point where $V(\Phi_a)$ is steepest, the sensitivity is maximum and can be approx-imated by the linearization of the $V(\Phi_a)$ characteristics. A change of flux by $\Delta\Phi_a = \Phi_0/2$, causes a change in current $\Delta I_{circ} = \Delta\Phi_a/L_{sq} = \Delta V/R_{sq}$. Consequently, the sensitivity is given by

$$V_\Phi = \frac{\partial V}{\partial \Phi_a} = \frac{R_{sq}}{L_{sq}} = \frac{R_{1,2}}{2L_{sq}}. \tag{2.56}$$

This means, that for highly sensitive SQUIDs, the inductance needs to be small and the resistance large. However, there are obvious limitations to both parameters. Particularly, the resistances $R_{1,2}$ still need to be sufficiently small, such that $\beta_c \leq 1$. One common approach is the miniaturization of SQUIDs, reducing the inductances $L_{1,2}$ and increasing the junction resistances, due to smaller dimensions. Another very promising approach is the introduction of asymmetries in the SQUID design, causing asymmetric $V(\Phi_a)$ dependences with one shallow side and one correspondingly steeper side [13, 71, 72][RNM+12]. The asymmetry parameters are typically denoted as α_I for the critical current, α_R for the resistance, α_L for the inductance and α_C for the capacitance ($\alpha \in [-1, 1]$). The SQUID parameters can then be expressed by

$$
\begin{aligned}
I_{c,i} &= I_0 \left(1 \mp \alpha_I\right), \\
R_i &= \frac{R_{sq}}{(1 \mp \alpha_R)}, \\
L_i &= \frac{L_{sq}}{2} \left(1 \mp \alpha_L\right), \\
C_i &= C_{sq} \left(1 \mp \alpha_C\right),
\end{aligned}
\tag{2.57}
$$

where $i = 1$ or 2, refers to the different junctions and accordingly to $\mp$. These asymmetries can be employed to maximize the transfer function V_Φ. However, for optimization of the overall SQUID performance with respect to the energy resolution, the effects of noise have to be taken into account [RNM+12]. A more detailed discussion will follow in subsec-tion 3.4.1, describing the optimized SQUID devices fabricated and investigated within this thesis.

2.4.2. Fractional Vortex Devices

Long Josephson junctions, exceeding λ_J in one lateral dimension, are essentially 1D systems for fluxons. As mentioned above, these fluxons are upheld by a circulating supercurrent across the tunneling barrier and carry the magnetic flux quantum due to flux quantization. In 1977 Bulaevskiĭ *et al.* proposed a system, having a ground state with a current and magnetic flux unequal zero, which consists of a superconducting ring interrupted by one Josephson junction with magnetic impurities in the barrier material [34]. Such magnetic impurities, cause a π phase shift of the Josephson phase [34]. When implemented in a superconducting ring, the ground state of the system is a non-zero supercurrent, compensating the artificial π-phase shift. Thus, the resulting magnetic flux through the ring is $\Phi_0/2$. Experimentally, such π-JJs can be realized by ferromagnetic barriers [73, 74], ramp type junctions consisting of s-wave and d-wave superconductors [75, 76], or grain boundary JJs between two differently oriented cuperate superconductors [77].

In a long junction consisting of alternating $0-$ and $\pi-$facets, half integer fluxons spontaneously appear at the $0 - \pi$-interfaces. These half integer fluxons are called semifluxons (SFs) and have been studied extensively over the last decades [16, 75, 78–80]. Unlike fluxons, semifluxons are pinned to the $0 - \pi$ boundaries and are consequently not solitons. They represent the ground state of a LJJ with π phase discontinuities and exhibit various highly interesting characteristics. The blue dashed line in Fig. 2.13 depicts (a) the phase, (b) the magnetic field and (c) the supercurrent of a semifluxon situated at $z/\lambda_J = 0$. For comparison, the corresponding characteristics of an integer fluxon are shown as a dotted red line. It can be seen that the phase discontinuity causes a circulating current over the tunneling barrier, resulting in either a direct semifluxon with $\Phi_{dSF} = \Phi_0/2$, or a complementary anti-semifluxon with $\Phi_{cSF} = -\Phi_0/2$, depending on the sign of the phase discontinuity $\pm\pi$. Accordingly, semifluxons behave like spin-($1/2$) particles with the two degenerate states $\pm\Phi_0/2$. Furthermore, they may be attributed with a mass and consequently exhibit an eigenfrequency. SFs of identical sign repel, while SFs of opposite sign attract each other [81]. Multiple semifluxons placed along a LJJ separated by a distance a, typically arrange antiferromagnetically, however they may also be polarized by external forces such as the bias current [82]. They can be excited using microwaves, creating an energy-band structure, which depends on the number of discontinuity points and the distance between two neighboring SFs [69].

Mathematically semifluxons can be described by a discontinuity of the Josephson phase φ at $z/\lambda_J = 0$, as shown in Fig. 2.13 (a). The step from a $0-$ to $\pi-$facet can be expressed by $\theta(\bar{z}) = \kappa H(\bar{z})$, where $H(\bar{z})$ is the Heaviside function and $\kappa = \pi$. In the perturbed Sine-Gordon equation this step is accounted for by an additional term $\theta_{\bar{z}\bar{z}}$, resulting in:

$$\varphi_{\bar{z}\bar{z}} - \varphi_{\bar{t}\bar{t}} - \sin\varphi = -\gamma + \alpha\varphi_{\bar{t}} - \theta_{\bar{z}\bar{z}}. \tag{2.58}$$

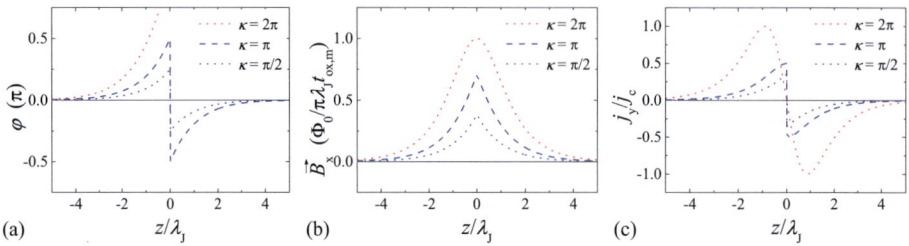

Figure 2.13.: (a) Phase response, (b) magnetic field and (c) supercurrent of fractional Josephson vortices. For comparison the dotted red line is depicted for the case of a full fluxon. The dashed blue line is for a semifluxon and the dotted blue line represents a fractional vortex with $\kappa = \pi/2$.

More recently, in 2004, a system based solely on conventional 0-JJs was proposed, allowing the creation of arbitrary fractions of fluxons [66]. The system effectively approximates $\theta_{\bar{z}\bar{z}}$ by attaching additional current injectors to one of the junction's electrodes. A comparable effect to that of the step function $\theta(\bar{z})$ can be achieved by applying a dc current to one of these injectors and extracting it from the other, resulting in a step-like function with the step height κ proportional to the injector current I_{inj}. Fig. 2.14 shows $\theta(\bar{z})$, its first and second derivative with respect to the coordinate $\bar{z}$ and the current profile $\gamma_\theta(\bar{z})$ fed to the junction using these current feed lines. Obviously $\gamma_\theta(\bar{z})$ is not equivalent to $\theta_{\bar{z}\bar{z}}(\bar{z})$, and therefore the finite width of the injectors W_{inj} and the distance between them dZ needs to be taken into account in simulation, design and experiment [83]. Where an injector pair having a width $W_{pair} = 2W_{inj} + dZ$ is small as compared to the Josephson penetration depth λ_J, the change of φ by κ can be assumed to be step-like. Ideally, the ratio W_{pair}/λ_J should be as small as possible, however in in experiment a ratio $\lesssim 3-4$ has proven to be sufficient. The striking advantage of these structures over the above mentioned $0-\pi$ JJs is the possibility of altering the strength of the phase-discontinuity by applying a dc injector current. By doing so, any arbitrary fraction of the magnetic flux quantum may be induced at the point of the injector pair. Hence, these vortices are called fractional or $\wp$-vortices. Furthermore, the complicated technology, combining conventional barrier materials with ferromagnetic materials,

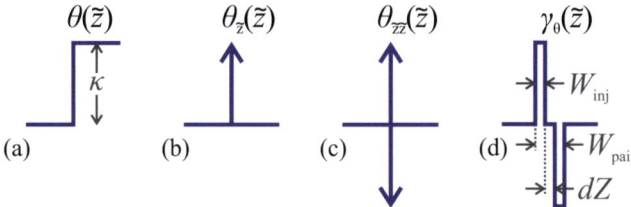

Figure 2.14.: (a) Step function $\theta(\bar{z})$ of height κ, (b) first derivative with respect to $\bar{z}$, (c) second derivative with respect to $\bar{z}$ and (d) and current profile $\gamma_\theta(\bar{z})$ of an artificial phase discontinuity.

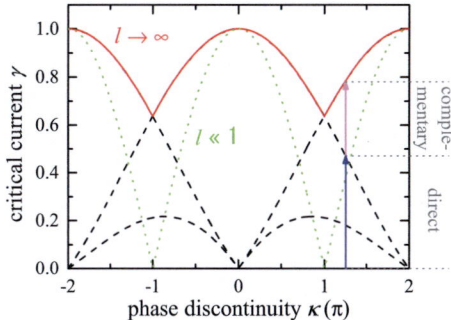

Figure 2.15.: Theoretical $\gamma(\kappa)$ dependence for an ideally abrupt phase discontinuity κ. The solid red line depicts the case of an infinitely long linear JJ, whereas the dotted green line shows the case for a short junction with $l \ll 1$. The dashed black lines show Fraunhofer patterns shifted by -2π and 2π.

or s-wave and d-wave superconductors, can be avoided. Instead, fractional vortex structures can be fabricated in the well-known Nb/Al-AlO$_x$/Nb technology. Fig. 2.13 shows (a) phase response, (b) magnetic field and (c) supercurrent of fractional Josephson vortices carrying $\Phi_0/4$ as a dotted blue line.

The magnetic component $\mu(\tilde{z}, \tilde{t})$ of the Josephson phase is continuous and related to the Josephson phase by $\varphi(\tilde{z}, \tilde{t}) = \mu(\tilde{z}, \tilde{t}) + \theta(\tilde{z})$ [16]. In an infinitely long JJ with a phase discontinuity κ at $z/\lambda_J = 0$ and no additional fluxons inside the junction, the topological charge of the fractional vortex can be expressed by $\wp = \mu(+\infty) - \mu(-\infty)$ [84]. For a direct $\wp$-vortex this results in $\wp = -\kappa$ and for its complementary counterpart $\wp = -\kappa + 2\pi \operatorname{sgn}(\kappa)$. A semifluxon therefore carries a topological charge π and a freely moving fluxon 2π. Vortex states are only stable for $|\wp| \leq 2\pi$ [85]. Just like semifluxons, $\wp$-vortices are pinned to their point of creation, *i.e.* the current injector pairs. Under small applied Lorentz forces they may shift slightly, but not move along the JJ. However, for a large enough bias current, the resulting Lorentz force may be strong enough to tear a full fluxon away, leaving a complementary $\wp$-vortex behind. For the special case of one fractional vortex in an annular LJJ, this depinning current has been derived [81, 86, 87] to be

$$\gamma_{c,\text{annular}}(\wp) = \left| \frac{\sin(\wp/2)}{\wp/2} \right|. \tag{2.59}$$

The maximum critical current of an annular *long* Josephson junction with one tunable phase discontinuity is therefore identical to the Fraunhofer pattern of a short junction described in section 2.2. It should be noted that this effect is independent of the JJ length.

For linear LJJs the depinning current shows a different behavior, due to the fact that full fluxons may escape the system at the ends of the junction. Thus, for infinitely long JJs ($l \to \infty$) Eq. (2.59) is only valid for $-\pi < \kappa < \pi$. For larger κ, the direct vortex changes to

Table 2.2.: Vortex notations typically used throughout literature and within this thesis.

Name	Symbol	Phase discontinuity	topological charge $\wp$
direct, up	$\uparrow$	$-\kappa$	κ
direct, down	$\downarrow$	$+\kappa$	$-\kappa$
complementary, up	$\Uparrow$	$+\kappa$	$-\kappa+2\pi$
complementary, down	$\Downarrow$	$-\kappa$	$\kappa-2\pi$
semifluxon	$\uparrow$	$\pm\pi$	π
antisemifluxon	$\downarrow$	$\pm\pi$	$-\pi$
fluxon	$\Uparrow$	0	2π
antifluxon	$\Downarrow$	0	-2π

a complementary vortex with a smaller topological charge once the bias current exceeds the corresponding value from Eq. (2.59). At this point a full fluxon is emitted from the LJJ. The total maximum critical current of an infinitely long linear junction is therefore described by the envelope of 2π shifted Fraunhofer patterns. Consequently the dependence is periodic with maxima at $\kappa = 2\pi n$ and cusp-like minima at $\kappa = (2n+1)\pi$. The modulation depth depends on the length of the JJ and is minimal for infinitely long JJs where $\gamma_{\min} = 2/\pi$. For JJs of finite length, the modulation depth increases and for short JJs ($l \ll 1$) γ modulates to 0, following $\gamma_{d,\text{short}}(\kappa) = |\cos(\kappa/2)|$, shown in Fig. 2.15 as a dotted green line. The solid red line in Fig. 2.15 shows the critical-current dependence on an applied injector current for an infinitely long linear JJ [83]. For $\kappa > \pi$ the vortex flips from a direct to a complementary vortex upon crossing a dashed black line corresponding to the 2π shifted, neighboring Fraunhofer patterns. In Fig. 2.15 this effect is indicated for $\kappa = 5/4 \cdot \Phi_0$. Tab. 2.2 summarizes the typical notations used for the different vortices described above.

Within this thesis various fractional vortex devices have been designed, fabricated and measured. Besides linear and annular LJJs with single phase discontinuities [85, 88][Bue13, Mer12][MMB$^+$13], vortex molecules with two $\wp$-vortices have also been investigated [Bue11][KMB$^+$12]. Typically, the devices have been dc characterized at the *Institute of Micro- and Nanoelectronic Systems* (IMS) at the *Karlsruhe Institute of Technology* (KIT), before being analyzed in more detail at the *Institut für Experimentalphysik II* (PIT II) at the *Universität Tübingen*. The performed experiments and measurement setups will be discussed further in subsection 3.4.2.

2.4.3. Josephson Oscillators

Over the last few decades the sub-mm wavelength range has received much attention. The interest mainly stems from radio astronomy since much of the cosmic background radiation, containing information about the formation of stars and our early universe, is in this

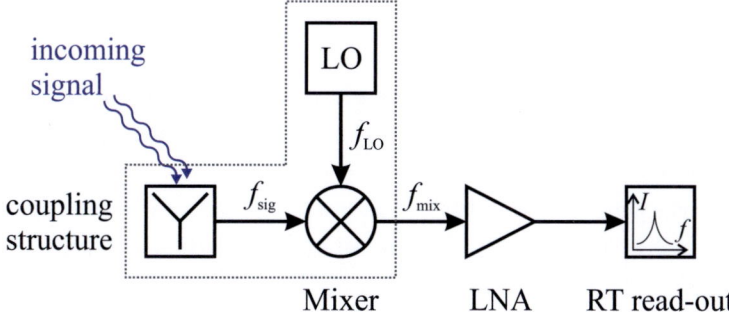

Figure 2.16.: Schematic representation of a heterodyne receiver system. The incoming signal is coupled to the receiver system via some sort of coupling structure, *e.g.* double slot antenna, and fed to the mixing element. In the mixer, the low-power signal frequency f_{sig} and the high-power local oscillator frequency are superposed, and the frequency $f_{\text{mix}} = |nf_{\text{sig}} - mf_{\text{LO}}|$ is passed on to a low noise amplifier (LNA). The amplified signal f_{mix} can then be post processed by conventional room temperature electronics.

electro-magnetic spectrum. However, this spectrum has been largely unexplored so far, mostly because of the strong absorption of THz radiation by water and water vapor in the earth's atmosphere. Besides the application in astronomy [89], there is also a large interest in the sub-mm spectroscopy in medical [90] and security applications [91]. All of these applications require low noise detector devices, allowing for high resolution spectroscopy.

There exist two principle detection mechanisms: direct detection and heterodyne detection. Detailed discussions of the different detection mechanisms exist in literature [92, 93]. This thesis included work on different detector types [Mar13][KSS$^+$13], however the focus was on heterodyne detector systems based on Josephson junctions.

Heterodyne receiver systems, as shown schematically in Fig. 2.16, are based on a superposition of the weak incoming signal to be measured and a strong signal generated by a local oscillator (LO). Both signals of frequency f_{sig} and f_{LO}, respectively, are fed to a mixing element with a non-linear IVC. Due to the non-linearity, the superpositioned signals cause frequencies

$$f_{\text{mix}} = |nf_{\text{sig}} - mf_{\text{LO}}|, \tag{2.60}$$

where n and m are integers. Typically, the frequency for $n = m = 1$ is the one of interest and will be referred to as the intermediate frequency $f_{\text{IF}} = |f_{\text{sig}} - f_{\text{LO}}|$. When f_{LO} is chosen sufficiently close to f_{sig}, the intermediate frequency is low enough (GHz range) to be processed by conventional room temperature electronics.

There are various devices available for the use as a mixing element. Schottky diodes are often used as a room temperature (RT) mixer. The fact that they can be operated at RT is the clear advantage over superconducting mixers. However, due to their limited non-linearity, their sensitivity is also limited. Furthermore, the required power coming from the local oscillator is large as compared to the cryogenic counterparts, such as the hot-electron bolometer (HEB) [94, 95] or the SIS mixer [96–98]. Using these highly sensitive HEB or SIS mixing elements also allows the integration of a local oscillator such as a FFO on chip. In this thesis, SIS harmonic mixers have been used as mixing elements for the investigation of improved FFOs. Tucker and Feldman [92] have found the optimal $\omega_{pl}RC$ product for SIS detectors to be in the range $3 - 4$, which has been confirmed in many experiments since [11, 12]. From the point of view of fabrication, this means that the junction should exhibit a high critical-current density and a large value of R_{sg}/R_N, while having a small capacitance at the same time.

The third focus of this thesis has been the development of an improved on-chip local oscillator for sub-mm receiver devices based on Nb/Al-AlO$_x$/Nb technology. In 1983 Nagatsuma *et al.* [8] proposed a linear long Josephson junction as a local oscillator and developed a complete device in the following years, suitable to deliver enough power ($\sim 10^{-6}$ W) for a mixing element such as a SIS or a HEB mixer [9, 10, 99]. In order to achieve such a high output power, the absolute critical current of the FFO needs to be sufficiently large. Thus a high j_c and a large junction area is beneficial for the operation of a LJJ as a FFO. In order to achieve high output frequencies, the energy gap of the superconducting electrodes should be as large as possible. The maximum frequency of a niobium based FFO is therefore given by $f_{g,Nb} \approx 677$ GHz. Essentially, the FFO consists of a LJJ with a magnetic field applied by a current I_{cl} passing in one of the electrodes, such that the junction is in the flux-flow regime. This current path is commonly referred to as the control line. In the flux-flow regime the emitted frequency f_{LO} can be tuned continuously up to f_g by the bias and control-line current, making the flux-flow regime the preferred operation region. Operation in the Fiske regime is also possible, however more complicated [100]. Regardless of the operation mode and the mixing element, the stability and linewidth of the frequency emitted by the FFO is a key aspect for the overall quality of the complete receiver device.

Fig. 2.17 (a) shows a schematic top-view of a FFO with a bias current I_b applied to the LJJ and a control-line current I_{cl} passing in the bottom electrode, perpendicular to the bias current. The top electrode is indicated in transparent light blue, the bottom electrode in dark blue and the junction area in red. I_{cl} creates a magnetic field through the barrier, indicated by green arrows, causing Josephson vortices to enter the LJJ, depicted as gray ellipses in Fig. 2.17 (b). These fluxons are accelerated in negative z-direction, due to the Lorentz force F_L created by I_b. At the end of the FFO these fluxons leave the LJJ, emitting electromagnetic radiation, which is then coupled to the mixing elements through an rf matching circuitry.

From the second Josephson equation it is known that the emitted frequency is proportional to the voltage across the JJ, *i.e.* the FFO is a voltage controlled oscillator (VCO). With typical dimensions of $L \approx 300\,\mu$m and $W \approx 10\,\mu$m, the normal resistance of the FFO fabricated in the Nb/Al-AlO$_x$/Nb technology is in the range of $1\,\Omega$ and is therefore usually current biased. Thus, if the bias current is afflicted with noise, automatically the voltage is

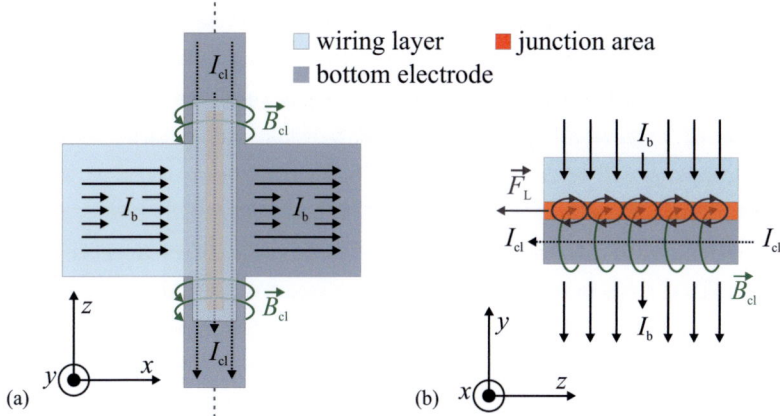

Figure 2.17.: Schematic of a FFO with applied bias I_b and control-line I_{cl} current. (a) Shows the top view of the FFO. (b) Shows the cross-section of the FFO along the z-axis indicated in (a) by a dashed gray line. The fluxons created by the magnetic field of the control-line current, experience a Lorentz force F_L in negative z-direction.

afflicted with noise. Additionally, the control-line current will introduce noise to the FFO. As is known from the theory of VCOs, the amplitude and spectral distribution of the voltage fluctuations will result in a broadening of the output linewidth Δf of the VCO. In case of a FFO, small changes in the bias current ∂I_b and control-line current ∂I_{cl} will result in voltage fluctuations given by:

$$\partial V_b = R_d \cdot \partial I_b + R_d^{cl} \cdot \partial I_{cl}. \tag{2.61}$$

Here, R_d is the differential resistance of the FFO with respect to changes in the bias current and R_d^{cl} with respect to changes in the control-line current. Consequently, fluctuations of the bias voltage ∂V_b, will result in an increase of Δf [89]. Within this thesis, FFOs have been investigated, where the bias current and the control-line current were correlated artificially, using an LC-shunt, such that the total fluctuations are somewhat compensated, leading to a reduced Δf. Linewidth measurements of conventional FFOs and LC-shunted FFOs have been done on devices integrated on chip with SIS mixing elements. The components, enclosed by the dashed gray line in Fig. 2.17 have been integrated on one Nb/Al-AlO$_x$/Nb chip. Additionally, the transmission of the microwave radiation from the FFO to the SIS mixer has been investigated. The obtained results will be discussed in subsection 3.4.5.

2.4.4. Requirements for Future Josephson Devices

In the previous subsections 2.4.1 - 2.4.3, Josephson devices that have been investigated within this thesis have been introduced. Many of the junction characteristics such as the I_cR_N product, the gap voltage and thus also the gap frequency, can be enhanced by changing the material composition of the trilayers. Already in the 1980's, first publications showed results on Josephson junctions based on niobium nitride instead of pure niobium [101–104]. In combination with aluminum nitride as a barrier material, NbN/AlN/NbN Josephson junctions can be fabricated, having superior characteristics as compared to Nb/Al-AlO$_x$/Nb JJs. By means of the previously introduced flux-flow oscillator, the restrictions of the Nb/Al-AlO$_x$/Nb technology will be discussed and possible improvements based on NbN and AlN will be introduced. Over the last few decades, various integrated receiver devices have been developed [89, 100, 105] and already several projects have been launched successfully where these devices are in use [106–109]. When based on Nb/Al-AlO$_x$/Nb technology, these receiver devices have an upper frequency limit inherent to the niobium SIS junction technology.

- The first restriction stems from the energy gap of the superconducting junction electrodes and the microwave transmission line. For the detector element, the theoretical frequency limit is given by $f_{\mathrm{lim}} = 4\Delta/h \approx 1350\,\mathrm{GHz}$ in case of a gap voltage of $V_{\mathrm{g,Nb}} = 2.8\,\mathrm{mV}$ ($\Delta_{\mathrm{Nb}} = 1.4\,\mathrm{meV}$). In practice, the surface losses increase rapidly above the gap-frequency $f_{\mathrm{g}} = 2\Delta/h$, leading to an effective upper frequency limit of the described Nb/Al-AlO$_x$/Nb technology of $f_{\mathrm{lim}} \approx 677\,\mathrm{GHz}$ [89].

- A second restriction arises from the SIS junction topography, representing a parallel plate capacitor, consisting of two superconducting electrodes separated by a very thin insulation layer. This inherent capacitance needs to be compensated by the on-chip rf matching circuit, which becomes increasingly complicated with rising values of the capacitance.

From these two restrictions, one can conclude that a technology allowing high critical-current densities, while maintaining a high junction quality to reach the optimal $\omega_{\mathrm{pl}}RC$ product, and a superconducting electrode material with a large energy gap in order to reach higher mixing frequencies, is desirable.

Addressing this, a multi-chamber *in-situ* sputter system for NbN/AlN/NbN technology has been developed within this thesis. Niobium nitride has approximately twice the energy gap of niobium and therefore theoretically allows mixing frequencies up to $f_{\mathrm{g,NbN}} \approx 1.4\,\mathrm{THz}$. Furthermore, using aluminum nitride instead of aluminum oxide as a barrier material yields a significantly lower specific capacitance, relaxing the restrictions concerning the inherent junction capacitance. This is particularly beneficial for high-frequency devices. In chapter 4 this system will be described and the results of the optimization of the NbN and AlN thin films, as well as the resulting NbN/AlN/NbN Josephson junction technology will be discussed in detail.

Conclusion

In this section the Josephson devices developed in the framework of this thesis have been introduced, starting with the theory on Superconducting Quantum Interference Devices. Two types of SQUIDs have been addressed in particular. First, the current-asymmetric 3-JJ SQUID for the read-out of different states of fractional vortices and second, the conventional dc SQUID based on 2 Josephson junctions. For the latter, asymmetries aside from the previously discussed current-asymmetry have also been introduced, leaving the possibility of maximizing the energy resolution somewhat.

In the second part of this section, the focus has been on fractional vortex devices based on long Josephson junctions. By means of the perturbed Sine-Gordon equation, the possibility of artificially creating vortices, carrying arbitrary fractions of the magnetic flux quantum Φ_0, has been derived.

The third Josephson junction device introduced is the Josephson oscillator. Specifically, so-called flux-flow oscillators based on LJJs, their working principle as well as their internal noise sources have been discussed. Additionally new ways of optimizing such FFOs have been addressed.

Finally, the section closes with a short summary of the restrictions of the Nb/Al-AlO$_x$/Nb technology. By means of the FFO specific high-frequency limitations have been addressed and a new NbN/AlN/NbN technology has been introduced to circumvent such restrictions. This technology has superior characteristics as compared to the technology based on niobium and aluminum oxide.

3. Josephson Junctions Based on Trilayers of Niobium and Aluminum Oxide

The following chapter focuses on the different fabrication processes developed and refined during this thesis. In the first section, the conventional Nb/Al-AlO$_x$/Nb process will be introduced, which was the starting point for this work. Based on this process, more sophisticated technologies have been derived and refined over the last few years including an Al-Hard-Mask process [KMI$^+$11]. In section two, a new fabrication process will be discussed. It has been developed to meet the requirements for high-quality Josephson junctions with minimized lateral dimension in all layers and exhibiting an almost planar surface of the final chips [MMB$^+$13]. The last section discusses the fabricated Josephson devices based on the Nb/Al-AlO$_x$/Nb technology and covers the results obtained from experiments on the devices introduced in section 2.4. Details on the used lithography techniques and the deposition and etching parameters are summarized in Appendix A.1 and A.2, respectively.

3.1. Conventional Fabrication Process

In the following section the conventional fabrication process for Nb/Al-AlO$_x$/Nb Josephson junctions will be introduced. It represents the starting point for the development of new fabrication processes within this thesis. In the beginning the deposition of the trilayer structures is described in detail, since it forms the common ground for all the Nb/Al-AlO$_x$/Nb based Josephson junctions discussed in the following. The advantages and drawbacks of the conventional process will be discussed in detail. Furthermore, the applicability of this particular process for electron-beam lithography and the achievable minimum lateral dimensions will be outlined.

As mentioned above all Nb/Al-AlO$_x$/Nb Josephson junctions discussed in this thesis are fabricated on an oxidized 2″ Silicon wafer diced in $10 \times 10\,\mathrm{mm}^2$ chips. Typically, oxidized low resistive Silicon in [100] orientation is used, with an average wafer thickness of 275 μm. The oxidation is performed in a temper oven, at a temperature $T_{ox} \approx 1000\,°\mathrm{C}$ for 4 hours and 100 % humidity. The thickness of the resulting amorphous SiO$_2$ layer is in the range of approximately 600 nm.

Trilayers of niobium and aluminum oxide are deposited at room temperature on the oxidized wafers in a high vacuum 2-chamber system with two 3″ dc magnetrons installed in the recipient. The system is based on a commercial system from Leybold [110], however has been modified to meet the requirements for trilayer fabrication. The main modifications include an argon rf plasma pre-cleaning system in the load-lock and an electronically controlled oxidation process. For the trilayer deposition, the wafer is fixed on a sample holder of stainless steel with an indium foil on top, in order to ensure good thermal contact of the wafer to the thermal bath of the sample holder.

After a 10 min Argon rf plasma pre-cleaning in the load-lock, the sample is moved to the recipient, using a mechanical transfer stage equipped with a clamping system. The base pressure of the sputter chamber is usually in the range of $2 \cdot 10^{-5}$ Pa, individually evacuated using a turbo molecular pump. Using a separate mechanical transfer mechanism, the sample is placed underneath the $3''$ niobium magnetron. Before actual deposition of the bottom Nb electrode the target is cleaned by pre-sputtering for 3 minutes, while the sample is still protected using a pivoting shutter. The typical thickness of the polycrystalline Nb bottom electrode, shown in Fig. 3.1, ranges from $90 - 100$ nm, which is slightly larger than the London penetration depth $\lambda_{L,Nb} \approx 85$ nm found in literature [111–113] and extracted from interferometer measurements. The top row in Fig. 3.1 shows images taken with the in lens detector, giving high contrast, whereas the bottom row shows images taken with the secondary electron detector, resulting in images of lower contrast, however including information about the surface smoothness and material homogeneity. It can be seen that despite clearly polycrystalline structure of the film (in the in lens image), the shading of the secondary electron image indicates a smooth surface and homogeneous material composition.

Subsequently, the sample is moved to the $3''$ Al-target and again protected using the pivoting shutter during the 10 min pre-sputtering. Upon opening the shutter, an aluminum wetting layer, as proposed by Gurvitch *et al.* [42, 43], is deposited. On the basis of the publications by Imamura *et al.* [114, 115], a thickness of $6 - 7$ nm has been chosen, sufficient to level the surface of the underlying niobium for the later growth of the tunneling barrier, without introducing significant anomalies in the *IVC* of the junctions [115, 116].

The sample is then transferred back into the load-lock, where the oxidation of the Al layer is done. As mentioned above, a modification of the system allows an electronically controlled oxidation. An in-house programmed micro controller controls an electro-magnetic valve for the oxygen inlet [117]. The user enters the desired pressure and upon starting the oxidation sequence, the μ-controller opens the valve for a short time (< 2 s), which is required for the setting of the oxygen pressure p_{oxy}. The oxidation time t_{oxy}, necessary for any wanted AlO$_x$ thickness, is controlled by the user manually. The total oxygen exposure $E_{ox} = p_{oxy} \cdot t_{oxy}$ governs the growth of the tunneling barrier and therefore the critical-current density of the trilayer according to the proportionality

$$j_c \propto \left(p_{oxy} \cdot t_{oxy} \right)^{-b}, \tag{3.1}$$

where b is an empirical constant, specific for each individual deposition system. Typical values for b, found in literature [47, 118, 119], range from 0.38 to 0.5, coinciding well

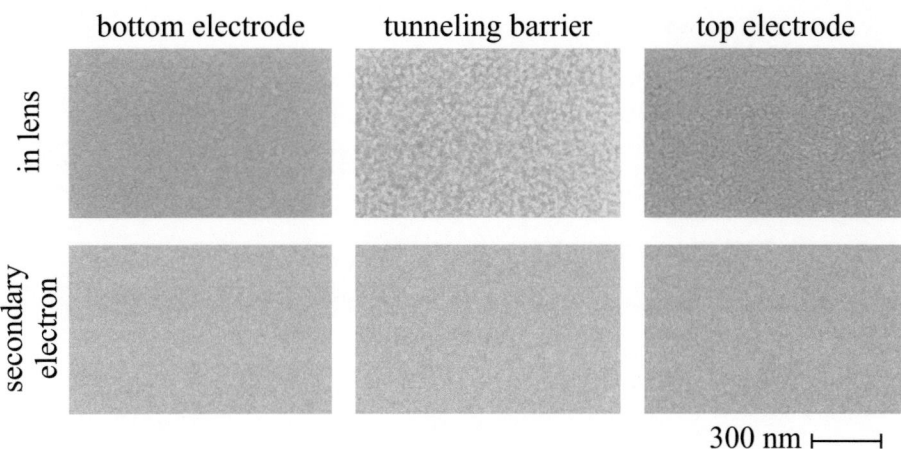

Figure 3.1.: Scanning electron microscope (SEM) images taken under 90° of (a) a 90 nm Nb film, typically used as the bottom electrode of a Nb/Al-AlO$_x$/Nb JJ, (b) the oxidized Al wetting layer, forming the tunnel barrier and (c) a 90 nm thick top electrode of a trilayer.

with the value extracted from the Nb/Al-AlO$_x$/Nb trilayers at the IMS, where $b = 0.48$ for critical-current densities $j_c \lesssim 15\,\text{kA/cm}^2$. For higher j_cs, the transparency of the barrier increases, leading to larger values of b [118]. The central images in Fig. 3.1 show scanning electron microscope (SEM) images of an oxidized 6 nm thick Al layer on top of the bottom Nb electrode. Again the in lens image shows some polycrystallinity, while the secondary electron image indicates a very smooth surface of the barrier layer.

After the oxidation, the sample is moved back into the main chamber, for deposition of the niobium counter electrode. This should take the least possible amount of time, in order to prevent any further unwanted oxidation or contamination of the tunneling barrier. The sputter procedure is identical to that of the bottom electrode. For the conventional process the thickness is typically $90 - 100\,\text{nm}$, however, reduced to 30 nm in some cases of a planar process. The right images in Fig. 3.1 show SEM images of a 90 nm thick top Nb electrode. The structure of the top electrode clearly differs from that of the bottom electrode, exhibiting larger grains. Nevertheless, the secondary electron image still shows that a homogeneous Nb surface with small surface anomalies is achieved. The completed trilayer is then covered with a protective layer of resist, before it is diced in $10 \times 10\,\text{mm}^2$ chips for further processing.

The fabrication of the JJ devices themselves begins with the patterning of the ground electrode. After thorough cleaning of the chip surface, using n-hexane, acetone and propanol, the chip is spin-coated with photoresist. The minimum lateral dimensions of this lithography step has been in the range of $3 - 4\,\mu\text{m}$ at the beginning of this thesis. Fig. 3.2 (a) shows a schematic cross-section of this first process step. Subsequent etching of the trilayer using reactive-ion etching (RIE), ion-beam etching (IBE) and a second RIE step, finalizes the

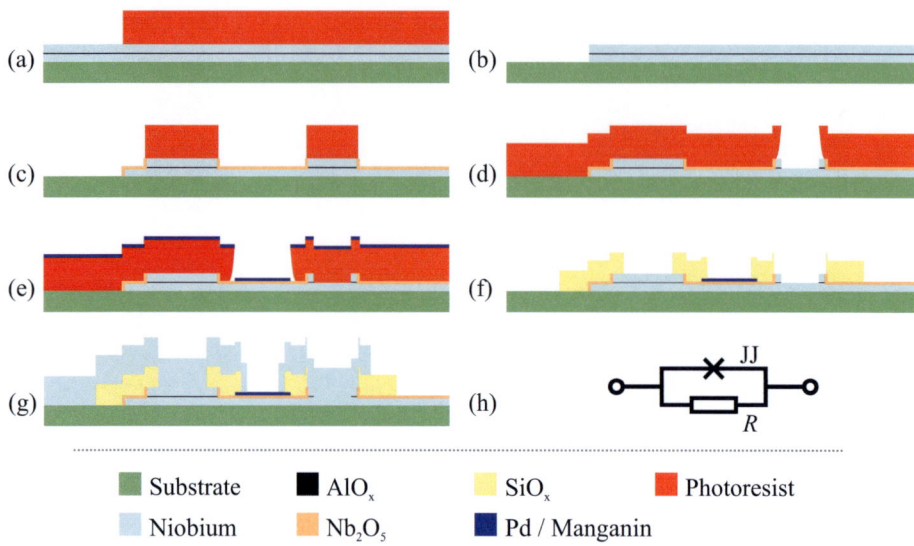

Substrate	AlO$_x$	SiO$_x$	Photoresist
Niobium	Nb$_2$O$_5$	Pd / Manganin	

Figure 3.2.: Schematic cross-sectional view of the conventional Josephson junction process. The individual steps from (a) to (g) are discussed in the text. The resulting chip topography in (g) exhibits multiple steps of maximum $350 - 450\,$nm height. The corresponding equivalent circuit diagram is shown in sub-figure (h).

ground electrode patterning. It should be noted, that a continuous etching through the entire trilayer is not possible, since the chemical share of the used RIE procedure does not etch Al or AlO$_x$ at all, and the mechanical impact, without an extra bias voltage, is not sufficient for physical etching. This makes the intermediate IBE etching step necessary, merely removing the barrier layer. Also it should be mentioned, that the selectivity between niobium and silicon dioxide during the second RIE is sufficient, so that neither an additional stopper layer, nor end-point detection is needed, whilst still maintaining an acceptably small over etching effect. Fig. 3.2 (b) shows a cross-section of the cleaned chip at this process step.

In Fig. 3.2 (c) the cross-section after the junction definition is shown. This step is again processed in a positive lithography step, leaving resist stencils for the junctions and vertical interconnect accesses (vias). Before the RIE etching of the top electrode the chip is post baked for 5 min at 120 °C, improving the resist's adhesion to the surface. This is necessary for obtaining a reproducible result in the subsequent anodic oxidation after RIE. The anodic oxidation forms a first insulating layer of Nb$_2$O$_5$ around the JJ definition, as depicted in Fig. 3.1 (c-f).

Next, the vias are etched, for which the resist is exposed using an image reversal lithography. This lithography technique leaves a very strong resist layer with an undercut as depicted in Fig. 3.2 (d). To open the vias, both the top electrode of the trilayer as well as the tunneling barrier is etched using a combination of RIE and IBE. After the definition of the vias, the individual structures on the chip are separated in an additional processing step consisting of positive lithography and consecutive RIE - IBE - RIE, which is not shown in Fig. 3.2.

For designs including resistor circuitry, *e.g.* overdamped junctions, palladium is usually used as a resistor material, exhibiting a sheet resistance of $1\,\Omega/\square$ at a thickness of $\sim 67\,$nm. Again, the patterning is done using the image reversal technique. The entire chip is then covered with the resistor material as depicted in Fig. 3.2 (e). After deposition, the remaining resist is dissolved in acetone, lifting the thin film off the chip. Accordingly, this technique is called lift-off. For this type of patterning, the mentioned undercut of the resist is particularly important, since lifting materials using a resist with positive edge slopes would inevitably result in torn edges. In case of sputtered material, this effect is very pronounced, due to the isotropic nature of the sputtering process. Furthermore, the resist layer should always be significantly thicker than the material to be lifted. One major drawback of palladium as a resistor material is its poor adhesion to Nb, SiO and SiO_2. Therefore, in the old conventional fabrication process, the resistor had to be placed on top of Nb_2O_5 being the only material available to which the adhesion is sufficient. To avoid shorts to the lower electrode, the anodization voltage had to be rather high for an electrically leak-proof insulation layer, *c.f.* Fig. 3.2 (e-g). With $0.88\,$nm Nb turned into $2.3\,$nm Nb_2O_5 per applied volt and a minimum Nb_2O_5 thickness of $\sim 40\,$nm the anodization voltage could not be lowered beneath $20\,$V. Unless the time consuming Al-Hard-Mask process is employed, this will always lead to a strongly reduced reproducibility for small JJs due to encroachment [KMI+11].

Next, the main insulation layer is deposited. At the IMS, silicon monoxide is thermally evaporated and again patterned using the lift-off technique. Due to the low base and deposition pressure of the system in the range of $10^{-5}\,$Pa, combined with the large distance between the crucible holding the granular SiO and the sample ($\sim 120\,$mm), the deposition is highly anisotropic. For a reasonably good edge coverage at the steps of the underlying chip topography ($\sim 90\,$nm), the SiO thickness should exceed this step height at least by $150 - 250\,$nm. The resulting SiO steps on the substrate surface is therefore in the range of $250 - 350\,$nm as shown in Fig. 3.2 (f).

The fabrication is finalized with the definition of the niobium wiring layer, see Fig. 3.2 (g). As for palladium, silicon monoxide and the vias, this lithography is also an image reversal step, and the niobium is structured using lift-off. For good superconducting connection between the surface of the trilayer's top electrode and the wiring layer, an *in-situ* 10 min pre-cleaning is performed. For good superconducting connection over the SiO steps, the thickness is normally in the range of $350 - 450\,$nm.

In Fig. 3.2 (h) the equivalent circuit diagram of a resistively shunted Josephson junction is shown, corresponding to the layer-by-layer description in the previous paragraphs. The minimum lateral junction dimensions in this conventional photolithography process are $\sim 4\,\mu$m. This is mostly due the patterning of the SiO insulation layer. The main drawback of lift-off is the limited stability of the resist stencil, which is strongly restricted by its lateral dimensions. For a $4 \times 4\,\mu m^2$ JJ, a SiO window of $2 \times 2\,\mu m^2$ needs to be aligned to the JJ. With the manual alignment stage available at the IMS photolithography system, this already poses the lower limit.

To overcome these restrictions, electron-beam lithography (EBL) can be employed for the JJ and SiO definition. In principle EBL allows lateral feature sizes in the sub-μm range, down to a few $10\,nm$ [120, 121]. To reach highest resolution, very thin resist ($\lesssim 100\,nm$) are necessary, which cannot be employed on uneven topologies with step heights that are in the same range as the resist thickness. Hence, in terms of typical EBL resist, rather thick resists have to be used, capable of withstanding RIE and anodic oxidation of the JJ definition, and suitable for lift-off of at least $250\,nm$ of SiO. For the junction definition a 3% solution of PMMA [122] resist is sufficient. Nevertheless, for sub-μm JJs, encroachment during the anodic oxidation underneath the resist stencil, led to poor reproducibility, since Nb_2O_5 was formed on the surface. To overcome this problem, an aluminum hard mask process was developed [47][KMI$^+$11], using patterned aluminum for the JJ definition, acting as an absolute etch stopper for RIE. During anodic oxidation, the Al is also anodized at a conversion rate of $1\,nm$ Al $\rightarrow$ $1.3\,nm$ of AlO_x per applied volt. The good adhesion of the Al to the Nb almost fully eliminates the encroachment during the anodization and thus improves reproducibility and allows for smaller JJ dimensions. After the anodization, the Al and AlO_x is removed in an aqueous 20%$_{vol.}$ solution of KOH.

Nevertheless, the main limitation of the minimum junction size still stems from the definition of SiO via, on top of the JJ. The vias through at least $250\,nm$ of SiO, still needs to be patterned using lift-off. The required resist thickness $\sim 500\,nm$ can only be achieved using a 5% solution of PMMA, which remains stable only for lateral dimensions comparable to its thickness. Consequently, SiO vias $d_{via} < 500\,nm$ could not be achieved reproducibly, limiting the minimal junction diameter to $d_{JJ} > 700\,nm$, assuming a $100\,nm$ overlap to account for the alignment accuracy.

Conclusion

In this section the conventional fabrication process, as developed prior to this thesis, has been introduced. A short description of the aluminum hard mask process outlined the possibilities of improving the reproducibility whilst reducing the lateral dimensions at the same time [47][KMI$^+$11]. The section ends with a summary of the limitations and drawbacks of the conventional and the hard mask process, which account for the need to develop a refined fabrication process, suitable for more sophisticated JJ devices.

3.2. Planar Fabrication Processes

For most applications of Josephson junctions, the minimal lateral dimensions of the conventional fabrication process are sufficiently small to meet design requirements. However, for designs with sub-µm features not only in the JJ definition, but also in the wiring layer, the real limitation stems from the uneven chip topography created during processing. Although, lateral dimensions of approximately 1 µm have been achieved using conventional photolithography at the IMS [123], the accuracy and reproducibility remains limited. For high-quality JJs with sub-µm feature sizes in all layers, a fabrication technology had to be developed, which allows the use of thin resist and therefore ensures maximum resolution. Consequently the chip topography had to be leveled after selected steps, creating flat surfaces. In this thesis such a planar fabrication process has been developed and will be described in the following. The requirements for fractional vortex structures designed to operate in the quantum regime will serve as an example.

As has been mentioned in section 2.2, the quantum regime is reached for temperatures below the crossover temperature given in Eq. (2.30). Since experiments become increasingly complicated with decreasing temperatures, it is desirable to reach the highest possible crossover temperature T^*. From Eq. (2.30) it can be seen that T^* is proportional to the critical-current density j_c of the trilayer. For a more detailed discussion of the exact critical current dependence, the reader is referred to [47] and the references therein. For simplicity, let us assume a proportionality $T^* \propto \sqrt{j_c}$, i.e. the critical-current density should be as high as possible. This led to the first requirement of an increasing j_c from approximately $600 \, \text{A/cm}^2$, at the beginning of this thesis, to values well above $1 \, \text{kA/cm}^2$.

While high critical-current densities are desirable to achieve high crossover temperatures and thus reduce complexity of the measurements, there also exists an upper limit of j_c, which stems from various aspects:

- First, j_c can only be increased when the thickness of the tunneling barrier is decreased, which results in an increasing transparency of the barrier and might also lead to a reduced sub-gap resistance [118, 119]. This inevitably results in a stronger damping of the junction, i.e. a drop in the R_{sg}/R_N-ratio. The limit is reached for a $R_{sg}/R_N = 10$, below which the damping is assumed to be too strong to be negligible for the investigation of vortex dynamics.

- Second, the typical upper limit of the LJJ's critical current is in the range of $\sim 1 \, \text{mA}$ for measurements at mK temperatures [124]. This limitation stems from the cooling power available at the mK-stage in a dilution fridge, which is in the range of a few 10µW at $\sim 20 \, \text{mK}$ and a few 100µW at $\sim 100 - 300 \, \text{mK}$ [125].

- The third and most limiting aspect stems from the specific design of artificial phase discontinuities. In subsection 2.4.2 the requirement of small injector pairs as compared to the Josephson penetration depth was introduced. However, from Eq. (2.31) it is known that the $\lambda_J \propto (j_c)^{-1/2}$, conflicting with the first aspect. The increase of j_c for obtaining high T^*, soon makes sub-µm injectors necessary, in order to satisfy $W_{pair} \lesssim 1/4 \cdot \lambda_J$. Furthermore it has to be taken into account that for a one-dimensional

45

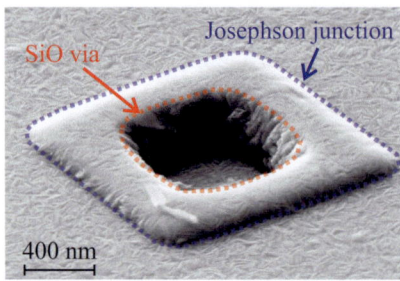

Figure 3.3.: SEM image of a junction fabricated using the Al-Hard-Mask process. Under the 60°
view, the large step heights are clearly visible, leading to the restriction concerning the
patterning of the Nb wiring layer.

flow of the injector current, the distance between the injectors should satisfy at least
$dZ \geq W$, the width of the junction. Thus, in order for $W_{pair} = 2W_{inj} + dZ$ to be sufficiently small, the junction also needs to be very narrow (sub-μm range) [83, 123].

In conclusion, the conventional fabrication process had to be refined with respect to the
range of the critical-current density, the lateral dimensions in the wiring layer (*i.e.* the width
of the current injectors) and consequently also with respect to the minimum junction geometry itself. In the following the development of a process, allowing for sub-μm dimensions,
is introduced, while the discussion of the larger range of critical-current densities can be
found in section 3.3.

As has been mentioned at the end of section 3.1, the conventional process creates a
strongly uneven chip topography and thus thick layers are necessary to ensure good edge
coverage. Fig. 3.3 shows a SEM image taken under a 60° tilt, clearly showing the large
step heights introduced during fabrication. The image shows the wiring layer covering
a $1.5 \times 1.5 \mu m^2$ JJ, processed using the Al-Hard-Mask process. Typically, the maximum
thickness of the available EBL resist is $\sim 500\,nm$. Patterning of the wiring layer using such
resists in a lift-off procedure is impossible in case of the uneven topography resulting from
the conventional process.

With growing complexity of the designs, an increasing number of layers becomes necessary, inherently resulting in a step-like topography of the chip-surface. There have been
many attempts to overcome these restrictions, including ramp-type junctions [126] and
chemical-mechanical polishing (CMP). While ramp type junctions are strongly restricted
in their lateral design and do not result in fully planar surfaces, CMP poses no limitations
in design nor surface-flatness. However, despite the fact that CMP would provide almost
perfectly planar surfaces, it will introduce a time-consuming fabrication step and cause mechanical stress to the junction, which may cause deterioration in the quality of the devices
susceptible to strain [127]. As a first attempt to circumvent this problem, a process based
on two separate insulation layers of SiO has been developed within this thesis, already re-

ducing the maximum step height by 50 %, as compared to that of the conventional process [Mer12]. This process was later expanded to include up to three insulation layers for the junction definition, reducing the step height down to a few 10 nm without the use of CMP or limitations in circuit design [MMB$^+$13]. This process is based on precise control of sputtering and etching rates of various materials and may be viewed as a combined derivation of self-aligned cross-type junctions [128, 129] and the above described conventional process. Whereas comparable processes [130] have been dismissed due to relatively low junction quality, thorough process refinements within this thesis have enabled fabrication of very high-quality junction devices. In the following, this refined process will be referred to as the self-planarized process.

Just like the conventional process described above, the self-planarized process is also based on processing previously deposited trilayers. The first step again consists of patterning of the ground electrode using RIE - IBE - RIE. However, instead of stripping the resist after the etching, the first self-aligned SiO insulation layer is deposited, as shown in Fig. 3.4 (a). Here the term self-aligned refers to a lift-off patterning step, using the identical resist, which has previously served as an etch-mask. For a self-aligned lift-off, the restrictions concerning the resist thickness are not as strict as for the conventional lift-off. This arises from the fact that the self-aligned material is deposited in a previously etched trench. Thus, even thin resist layers with thicknesses below the thickness of the later deposited material may be lifted, as long as the depth of the trench combined with the remaining resist after etching is thicker than the material to be lifted.

By this self-aligned deposition, an almost planar surface is created (see Fig. 3.4 (b)), permitting the use of thin resists for the junction and via definition, as indicated in Fig. 3.4 (c). During the etching of the top electrode, the first planarization layer is also exposed to the plasma. The selectivity between the two materials defines the exact thickness of the initial SiO layer. Once the top electrode is etched, the chip surface is anodized, growing an insulation layer of Nb_2O_5. Here, the anodization voltage is much lower as compared to the voltage applied in the conventional process. An adequate insulation is still reached, due to a subsequently evaporated second SiO layer. Depending on the top electrode's thickness, the SiO layer is usually 30 nm or 90 nm, leveling the topography.

Fig. 3.4 (d) depicts the cross-section of the chip, after preparation of the vias. This step includes a lithography for the areas where the via needs to be etched using RIE and IBE and subsequent sputtering of Nb. In order to ensure a high-quality via, a 10 min pre-cleaning is performed, before the niobium is sputtered and lifted. After cleaning the chip in acetone and propanol, the surface of the chip is almost ideally planar. Typical step heights measured using a profilometer ranged from ~ 10 nm down to immeasurably small steps.

The third insulation layer depicted in Fig. 3.4 (e) is patterned in a conventional lift-off process, and hence cannot be leveled. However, where the JJs and vias are defined using the strongly anisotropic ion-beam etching, the overetching effect is negligible, leaving almost

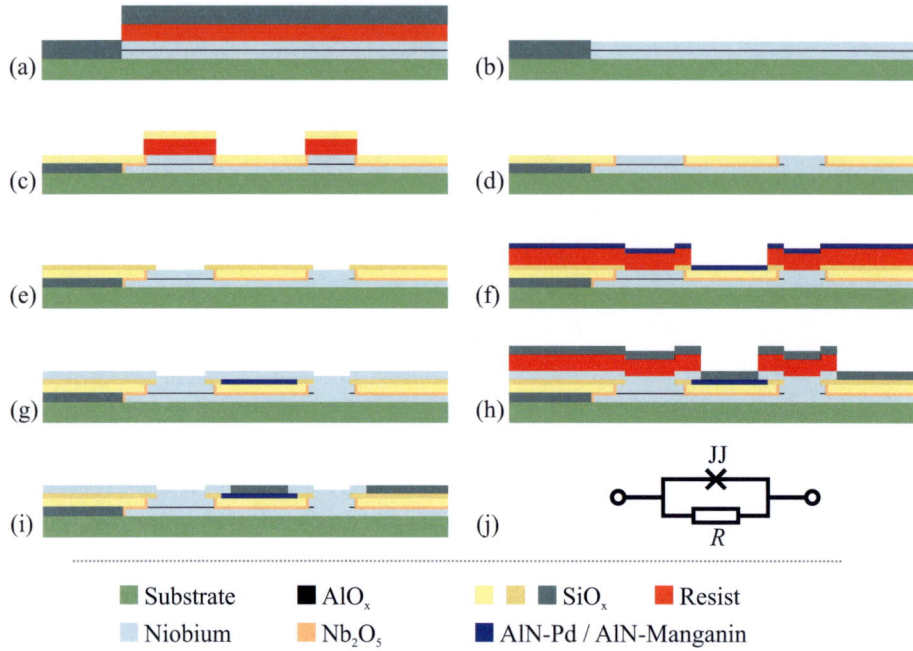

Figure 3.4.: Schematic cross-sectional view of the planar Josephson junction process. The individual steps from (a) to (i) are discussed in the text. Compared to the resulting chip topography of the conventional process, the self-planarized process exhibits much fewer and significantly smaller steps, *c.f.* sub-figure (i). Even with the optional third insulation layer, the maximum step height is merely a few 10 nm. The equivalent circuit diagram of a resistively shunted JJ is shown in sub-figure (j).

no trench between the Nb stencil and the surrounding self-aligned second SiO layer, making the third insulation layer obsolete. For the sake of completeness, this process step shall be described nonetheless. Originally the junction definition was done using RIE, causing a trench around the junction, which had to be compensated by this additional insulation layer. Typical thicknesses are in the range of 30 nm, defining the maximum step height of the entire chip topography. When comparing to the conventional process, this is already an improvement of approximately 90 %.

In Fig. 3.4 (f) the deposition of a shunt resistor is depicted. At this point it is important to note that the resistor may also be deposited in a planar fashion, by previously etching a shallow trench in the SiO layer. However, the planarization effect is limited by the thickness of the existing SiO, possibly making it necessary to reduce the thickness to values lower than the design value for $1\,\Omega/\square$ (~ 67 nm) and consequently resulting in higher sheet resistances. In most cases this effect can be compensated by the circuit design. Concerning the poor adhesion of palladium to SiO, various tests were performed in order to avoid this problem. Initially, manganin was used instead of palladium, exhibiting a stronger adhesion, but introducing a much higher sheet resistance than palladium. Although this allowed fabrication

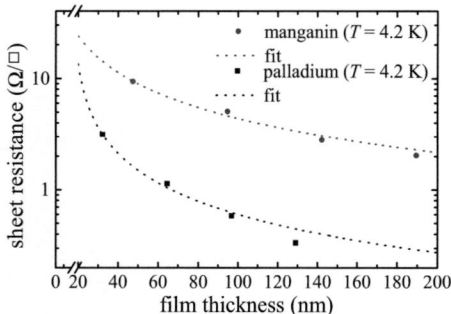

Figure 3.5.: Sheet resistances of palladium and manganin at $T = 4.2\,\mathrm{K}$. Both, palladium and manganin have been deposited on a 2 nm AlN buffer layer for improved adhesion to the underlying SiO$_2$.

of integrated circuits including resistors in the planar process, it strongly limits the range of sheet resistance. In order to continue the use of palladium a 2 nm buffer layer of aluminum nitride, was introduced as an adhesion layer underneath the palladium. Fig. 3.5 shows the sheet resistance of palladium and manganin on various thicknesses and the corresponding fits [131]. As can be seen, the sheet resistance is easily adjusted from $R_\square = 0.3 - 20\,\Omega/\square$.

Exhibiting a maximum step height in the range of only 30 nm, it is now possible to deposit a comparably thin niobium wiring layer of only ~ 150 nm, still achieving sufficient edge coverage and thus good superconducting properties across the entire chip, *c.f.* Fig. 3.4 (g). It should be noted that the lithography leaving the alignment marks at the chip's edges covered with resist before Nb is deposited is not shown in Fig. 3.4. Nevertheless, this is important for the alignment of the following lithography step depicted in Fig. 3.4 (h). Unlike in the conventional process, here the wiring layer is patterned by etching and optional subsequent SiO deposition. Due to the almost planar surface, a thin resist layer may be used, allowing for higher resolution lithography while still being resistant enough for etching. This procedure makes reliable photolithographic patterning down to 1 μm lateral dimensions possible.

In order to achieve even smaller dimensions, an electron-beam lithography process has also been developed for the planar fabrication process. Most chip designs, consist of a combination of large areas in the range of several 10 μm up to a few mm, and fine structures in the μm range of the JJs and the vicinity thereof. To minimize the turn-around time of the entire process including EBL steps, a so-called mix-and-match process has been developed, based on the resist AR-N 7520.18 [122]. This resist is sensitive to electron beam exposure as well as UV exposure, exhibiting a negative tone behavior, *i.e.* the exposed areas remain as resist stencils after development in the diluted standard MIF 300-47 developer. A combination of EBL and UV exposure thus allows to quickly pattern a standardized structure of bond pads and wide current feed lines using conventional photolithography, while the

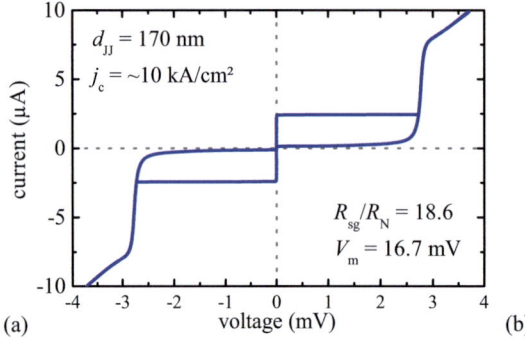
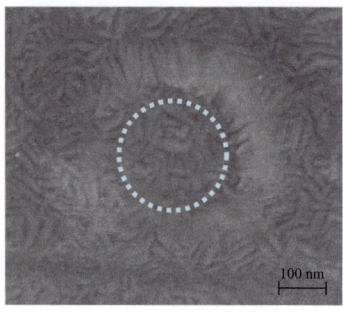

(a) (b)

Figure 3.6.: The *IVC* of a sub-μm JJ, measured at $T = 4.2\,\text{K}$ is shown in (a) and the corresponding SEM image taken under 90° is shown in (b). The suppressed I_c is due to the very small effective junction area of merely $A \approx 0.023\,\mu\text{m}^2$, which was estimated from the electrical properties. The design diameter $d_{\text{JJ}} = 200\,\text{nm}$ is indicated in (b).

fine JJ structures may be exposed by direct electron beam writing. Thorough adjustment of the lithography parameters, including the baking times, generates a resist mask, which is strong enough to withstand patterning of the ground electrode undergoing RIE - IBE - RIE and subsequent SiO deposition, while lift-off is still possible. Furthermore even 2:1 diluted resist, with 2 volume parts of resist to 1 part thinner, has proven to be resistant to 10 min of IBE for the JJ definition, despite its low thickness of merely 90 nm. This mix-and-match process can be employed in all layers of the fabrication process, hence enabling sub-μm structures defined by EBL. Minimum JJ sizes down to $\sim 200\,\text{nm}$ in diameter and injectors of only 600 nm in the wiring layer have been achieved [MMB$^+$13]. Fig. 3.6 (a) shows an *IVC* of a JJ with an estimated effective area of $A \approx 0.023\,\mu\text{m}^2$, having a critical-current density of $j_c \approx 10\,\text{kA/cm}^2$. Despite the extremely small lateral dimensions the junction exhibits very high quality parameters of $R_{\text{sg}}/R_{\text{N}} = 18.6$ and a characteristic voltage of $V_{\text{m}} = 16.7\,\text{mV}$. In Fig. 3.6 (b) the corresponding SEM image, taken under 90°, is shown and the design diameter of 200 nm is indicated by a dashed blue line. The effective junction diameter of $d_{\text{JJ}} \approx 170\,\text{nm}$ was estimated from the junction's electrical properties. The low contrast between the surrounding wiring layer and the junction area is due to the planar process without the optional insulation layer after the junction definition, resulting in an almost perfectly flat surface and thus leaving no steps.

In addition to the achievements in reducing all lateral and vertical dimensions of the fabrication process mentioned above, the integration of higher levels of wiring layer shall also be mentioned at this point. Even though the integration of a third metallization layer has previously been achieved using special kinds of very thick resist [Bue11], the minimum lateral dimensions of this third layer was limited to 3 μm and larger. Also this layer was purely passive, *i.e.* there were no via connections possible due to the very thick SiO insulation layer of 700 nm above the first wiring layer. Employing the new self-planarized

Table 3.1.: Comparison of the minimal feature sizes, that have been achieved using the conventional fabrication process and the newly developed self-planarized fabrication process. the acronym PL stand for photolithography and EBL for electron-beam lithography.

Layer	Short name	Conventional process		Self-planarized process	
		PL	EBL	PL	EBL
ground electrode	M2a	$4-5\,\mu m$	–	$\sim 3\,\mu m$	$\sim 1\,\mu m$
top electrode	M2b	$3-4\,\mu m$	$\sim 700\,nm^*$	$\sim 1.5\,\mu m$	$\sim 200\,nm$
insulation	I1b	$2-3\,\mu m$	$\sim 500\,nm^*$	$\sim 1\,\mu m$	$\sim 100\,nm$
wiring	M3	$\sim 1\,\mu m$	$\sim 1\,\mu m$	$\sim 1\,\mu m$	$\sim 600\,nm$
optional wiring layers follow below					
insulation	I2	$\sim 4\,\mu m$	–	$\sim 1\,\mu m$	$\sim 1\,\mu m$
wiring	M4	$3-4\,\mu m$	–	$\sim 1\,\mu m$	$\sim 200\,nm^{**}$

* Applies only when Al-Hard-Mask is employed [KMI+11].
** Value corresponds to the width of a slit in a coplanar waveguide design.

process, it is now easily possible to integrate higher levels of superconducting wiring, since the chip topography is very flat. Within this thesis, a third niobium layer has been integrated, connected to the ground electrode using vias [Bue13]. The patterning of this layer has been done in combination of conventional lift-off and subsequent positive photolithography and RIE. By this, minimum dimensions in the range of 200 nm have been achieved in a coplanar waveguide for coupling of microwaves to a LJJ. Tab. 3.1 shows a direct comparison of the minimum lateral feature sizes, which can be realized with the conventional and self-planarized fabrication process.

Conclusion

In order to meet the requirements for highly sophisticated Josephson devices with sub-μm lateral dimensions, the conventional fabrication process at the IMS was refined to a new self-planarized Josephson junction fabrication process. Self-aligned SiO layers, deposited after individual etching steps, level the chip's topography. This allows using thinner resists than in the conventional process, resulting in a higher resolution. Whereas the minimum junction size realized using photolithography in the conventional process was limited to $\sim 4 \times 4\,\mu m^2$, fabrication of junctions down to a diameter of approximately 1.5 μm is possible when using the refined process, while still exhibiting very good quality parameters [Mer12]. Using electron-beam lithography, it was possible to further reduce the minimum junction size from a diameter of approximately $d_{JJ} = 700$ nm (using the Al hard mask process) down to $d_{JJ} \approx 200\,\text{nm}$ [MMB$^+$13]. The higher resolution of the planar process comes at the cost of a longer turn-around time, which however, has been minimized by the development of a mix-and-match process combining EBL and UV lithography.

3.3. Extraction of Quality Parameters

This section is dedicated to the characterization of the employed fabrication processes. At first the measurement setup will be described and the changes as compared to the old setup will be discussed. The second part deals with the characterization of short Josephson junctions, which are shorter than the Josephson penetration depth introduced in Eq. (3.5). Typical lateral dimensions range from a few µm down to a few 100 nm. This subsection concludes with the measurements of serial arrays of short JJs, representing a possibility for probing the trilayer quality over a large distance of several mm.

Afterwards, a detailed discussion of how to probe the quality of extended Josephson junctions, significantly exceeding λ_J in one dimension, is presented. A new approach for probing junctions of intermediate length ($\approx \lambda_J$) up to very long JJs up to $30\lambda_J$ is introduced.

3.3.1. Measurement Setup

Measurements for basic characterization of short and long Josephson junctions have been performed at the IMS. The available setup includes a dipstick for measurements in liquid helium (LHe) in a transport dewar and is equipped with a single shield of cryoperm for shielding magnetic stray fields, such as the earth's magnetic field. This cryoperm sufficiently shields stray fields and is thus suitable for measurements of short and long Josephson junctions. Long junctions, significantly exceeding $\lambda_J \geq 10$, are however extremely susceptible to magnetic fields, often leading to trapped flux in the junction and therefore to suppressed critical currents and asymmetries in $I_c(\Phi)$ measurements. Consequently it is necessary to thermally cycle LJJs several times, to get flux-free states.

The dipstick design leaves the possibility of mounting different end-pieces designed for different measurements. Fig. 3.7 (a) shows different end-pieces for the JJ-setup along with a chip carrier. In (b) the electronic rack, the dipstick and a LHe dewar is shown. The setup includes 6 individual channels for 4-point measurements and an extra dc bias line for applying a magnetic field. There are two copper coils available, which can be mounted on the end of the dipstick. The larger one has 9195 windings and smaller one has 1080 windings. With the geometries indicated in Fig. 3.7 (c), this results in a magnetic fields of $0.162\,\text{mT/mA}$ and $0.047\,\text{mT/mA}$ for the large and small magnet respectively.

The measurements are carried out using the software *GoldExi* [132], which controls two PCI D/A- A/D-converter cards. The setup allows simultaneous controlling of up to ten individual current sources and recording of up to eight voltage responses of the device under test (DUT). Each of the current sources, as well as the voltage amplifiers, is battery powered

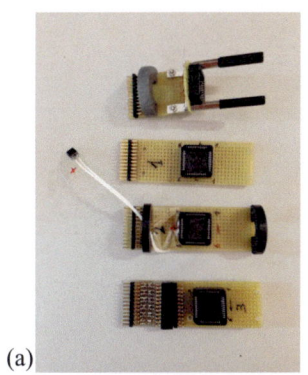 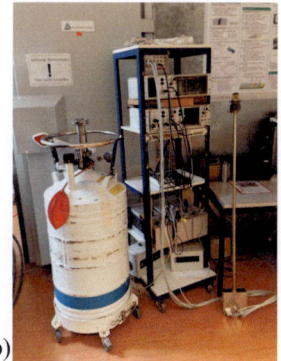 

(a) (b) (c)

Figure 3.7.: Josephson junction measurement setup at the IMS. (a) Shows different end-pieces for sample mounting and a chip carrier, (b) shows the electronics rack, the dipstick and a LHe dewar and (c) shows the two copper coils for the creation of magnetic fields.

individually. The current sources can be set to 5 different current ranges up to $\pm 0.1\,\text{mA}$, $\pm 1\,\text{mA}$, $\pm 10\,\text{mA}$, $\pm 100\,\text{mA}$ and $\pm 250\,\text{mA}$. This allows measurements of JJs in a large range of current densities.

All measurements discussed in the following have been performed at $T_{\text{LHe}} = 4.2\,\text{K}$, unless otherwise indicated. The fabricated JJ devices discussed in section 3.4, have each been measured in special setups, particularly designed for the specific measurement and will be described in the corresponding subsections.

3.3.2. Characterization of Short Josephson Junctions

For Josephson junctions, with lateral dimensions smaller than λ_J, the phase may be considered to be a constant value over the entire junction area and thus a point-like variable. Transport measurements of such JJs allows extraction of various quality parameters, *c.f.* section 2.2. For all fabrication processes described in the previous sections 3.1 and 3.2, these parameters are very high. However, it should be noted that most of the μm-sized JJs have been fabricated in the conventional process while the sub-μm sized JJs have been fabricated using the planar process employing electron-beam lithography. Since the most important aspect for high-quality Josephson junctions is a homogeneous tunneling barrier over the entire wafer, the fabrication processes will only be distinguished between if critical to the extracted information.

As mentioned before the critical-current density before the start of this thesis was limited to a range of approximately $50 - 600\,\text{A/cm}^2$. For devices such as SIS mixers and experiments in the quantum regime at $T < T^*$, higher j_cs are necessary. Usually the thermal oxidation of the aluminum layer is performed for $t_{\text{oxy}} = 30\,\text{min}$ with a minimum oxygen pressure of $p_{\text{oxy}} = 10\,\text{Pa}$. In order to increase the current density of the trilayers, the oxygen exposure $p_{\text{oxy}} \cdot t_{\text{oxy}}$ had to be reduced. From literature it is known that j_c is approximately

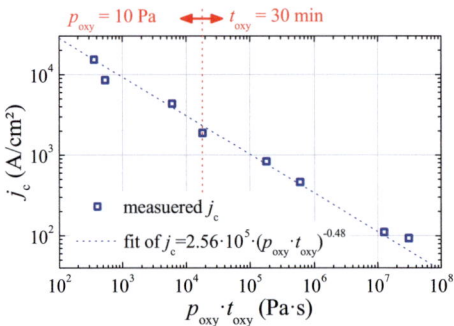

Figure 3.8.: Dependence of the critical-current density j_c on the oxidation atmosphere $p_{oxy} \cdot t_{oxy}$. The behavior is almost identical to the one found in [119].

proportional to $(p_{oxy} \cdot t_{oxy})^{-1/2}$ [119]. For the electronically controlled oxidation installed at the IMS, reduction of the oxidation pressure is only possible by changing the oxidation mode from a static atmosphere to an oxygen flow. The consequences for the trilayer quality of such a change are hard to predict, which is why it has been chosen to reduce the oxidation time instead of the pressure. Fig. 3.8 shows the dependence of the critical-current densities on the oxidation atmosphere extracted from representative IVC measurements. The found behavior, corresponds closely to the values from literature. The maximum measured j_c was 15.3 kA/cm^2 while the minimum was in the range of 50 A/cm^2. The two different oxidation regimes, the first with constant p_{oxy} and varying t_{oxy} and the second with varying p_{oxy} and constant t_{oxy}, are indicated in red.

For determination of the scalability of the junction sizes, a mask has been designed with various different JJ designs including 31 junctions of areas ranging from $2.25 - 81\,\mu m^2$. Fig. 3.9 gives an overview over typical quality parameters extracted from a representative chip fabricated using photolithography [Mer12]. In (a) the IVC of a 8.45 μm^2 JJ is depicted. The junction represents the typical quality of the Nb/Al-AlO$_x$/Nb JJs. All quality parameters, introduced in section 2.2, have been extracted from the measured IVCs as indicated in Fig. 2.4. The high gap voltage of $V_g = 2.82$ mV indicates that the niobium electrodes in vicinity of the barrier is of very high quality. The maximum V_g measured at the IMS for niobium JJs has been 2.95 mV which is close to the to the theoretical maximum of 2.96 mV at $T = 4.2$ K, indicating an almost perfect interface between the AlO$_x$ barrier and the Nb electrodes [133]. Additionally, the quality of the electrode material has been checked by $R(T)$ measurements and extraction of the critical temperature T_c as well as the residual resistance ratio RRR, which are typically 9.25 K and $\gtrsim 3$ for 100 nm thick films, respectively. From the measured critical current $I_c = 173.6\mu$A, a critical-current density of $j_c \approx 2.05$ kA/cm^2 can be calculated. The high value of the Ambegaokar-Baratoff parameter of $I_c R_N = 1.74$ mV, also indicates a good barrier interface.

As can be seen in Fig. 3.9 (b) the normal state resistance R_N scales almost ideally with

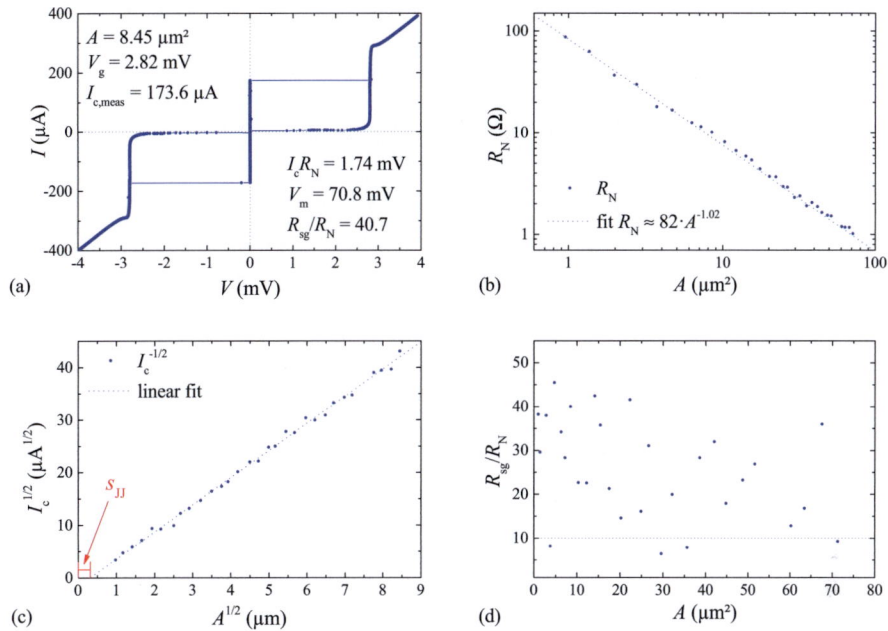

Figure 3.9.: Typical characteristics of junctions fabricated at the IMS [Mer12]. (a) Shows an *IVC* of a 8.45 μm² JJ of typical quality. (b) Shows the dependence of the normal state resistance on the junction area. In (c) the square root of the measured critical current is plotted over the square root of the junction area. The intersection of the linear fit with the abscissa indicates the linear shrinkage of the effective JJ size. (d) Shows the R_{sg}/R_N-ratio for all junctions shown in the previous graphs (a-c).

$1/A$, as expected from theory. Figure (c) shows the square root of the measured critical current over the square root of the junction area. The linear dependence is expected for junctions of different sizes from one trilayer. From the dependences shown in figures (b) and (c), it can be inferred that the $I_c R_N$-product stays constant over a wide range of junction area, only deviating towards smaller JJ sizes. This deviation is mainly due to the nonlinear shrinkage effect for JJs of decreasing size. The isotropic nature of reactive-ion etching for the JJ definition and the subsequent anodic oxidation causes the effective area of the junction to shrink as compared to the designed value. Additionally, the resist stencil for the JJ definition can deviate from the design on the chromium mask, due to over- or under-exposure and development during lithography. The lithographic effect is accounted for in Fig. 3.9 since the JJ areas have been measured using a scanning electron microscope. The intersection of the linear fit with the abscissa of figure (c) corresponds to the linear shrinkage of square junctions according to:

$$s_{JJ} = w_{des} - \sqrt{\frac{I_{c,sw}}{j_c}}, \qquad (3.2)$$

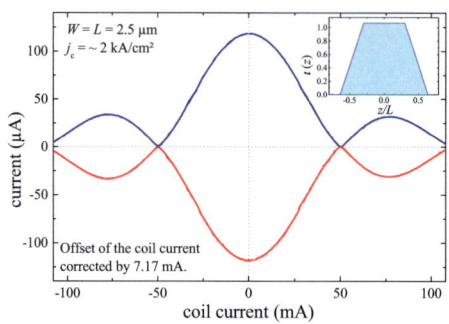

Figure 3.10.: $I_c(H)$ dependence of a square Josephson junction. The deviation from the Fraunhofer pattern, stems from a misalignment of the external magnetic field, effectively causing the current distribution to follow a trapezoidal shape, as indicated in the inset.

where $I_{c,sw}$ corresponds to the critical current extracted from the IVC and j_c to the average critical-current density of the wafer. w_{des} corresponds to the designed lateral dimensions W and L of the square JJs. For the shown set of junctions in Fig. 3.9 this shrinkage is approximately $s_{JJ} = 310$ nm.

In figure (d) the ratio between the sub-gap resistance R_{sg} and the normal state resistance R_N is shown. As can be seen, the ratio is above 10 for almost all measured JJs. Taking into account the high j_c of more than 2 kA/cm^2, values of up to 46 are extraordinarily high for Nb/Al-AlO$_x$/Nb junctions. The high quality of the barrier and the extremely low sub-gap leakage current is also evident from the high V_m values of up to 80.

From the modulation of the maximum critical current under the influence of an externally applied magnetic field, information on the barrier homogeneity can be inferred. Fig. 3.10 shows an $I_c(H)$ dependence of a $6.25\,\mu m^2$ junction with an approximate critical-current density of $j_c = 2$ kA/cm^2. The data has been recorded for positive and negative bias currents and has been corrected by an offset of 7.17 mA in the coil current. The perfect symmetry and the modulation down to $I_c = 0$ mA show the high quality of the trilayer. The fact that the modulation does not fit a Fraunhofer pattern, is due to a misalignment of the square junction with respect to the magnetic field lines. Thus, the magnetic field is not parallel to the x-direction as indicated in Fig. 2.5. Consequently, the $I_c(H)$ dependence corresponds to the Fourier-transform of a trapezoidal current distribution, as schematically indicated in the inset of Fig. 3.10 for a rotation of $20°$ relative to the magnetic field in the yz-plane.

Serial arrays of short junctions, spread over a large distance of a chip, were used to check the variation of the current distribution across a 10×10 mm^2 chip [Heu10]. The arrays of ~ 100 JJs were all on a length scale $\gtrsim 1 - 2$ mm and consisted of $7 \times 7\,\mu m^2$ or $10 \times 10\,\mu m^2$ JJs. Fig. 3.11 (a) shows a histogram of switching currents of a serial array of 100 junctions. The corresponding IVC is depicted in the inset. Despite the fact that the histogram is not corrected for variation of the actual lateral junction dimensions, the standard deviation of

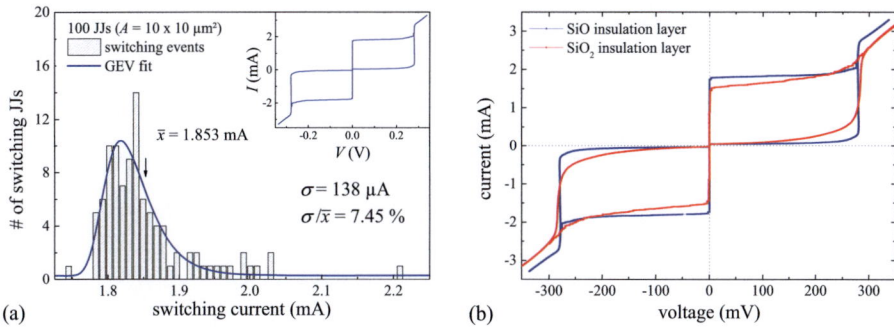

(a)   (b)

Figure 3.11.: Measurements of serial junction arrays. (a) Depicts a switching-current histogram of an array of 100 JJs. The *IVC* is shown in the inset. In (b) the effects of differently deposited insulation layers can be seen. Whereas the electron discharge during rf sputtering leads to significant degradation of the barrier, the leakage remains minimal for thermally evaporated SiO.

the recorded switching currents is merely $\sigma = 138\mu A$. With a mean critical-current density of $j_{c,mean} = 1.85\,kA/cm^2$, the array exhibits a coefficient of deviation of $\sigma/\bar{x} = 7.45\,\%$, where $\bar{x} = 1.853\,mA$ is the mean switching current of the array. Fig. 3.11 (b) shows two *IVC*s of 100-JJ-arrays, both fabricated on the identical trilayer in the conventional process, however with different insulation layers [134]. While the red trace depicts a chip with a silicon dioxide insulation layer, the blue trace corresponds to a chip with silicon monoxide insulation layer. It can clearly be seen that the switching-current spread for the SiO_2 trace is significantly larger than for the blue SiO trace. This effect can be explained by the different deposition procedures of the two materials. Whereas SiO_2 has been rf sputtered, leading to a bombardment of the chip with highly energetic electrons, SiO has been thermally evaporated and hence not exposed to any charge fluctuations. From these experiments it can be concluded that any sort of plasma discharge incident directly on chip, may lead to degradation of the JJ quality. The highest quality of junctions has been realized when the individual islands with junctions on a chip have been isolated from each other as early as possible during fabrication [Heu10], *i.e.* right after the anodic oxidation.

From these measurements, it can be concluded that the homogeneity, as well as the overall trilayer qualities are very good. Taking this homogeneity into account, the critical-current density extracted from *IVC* measurements can be re-evaluated. When plotting $j_c(A)$ a significant drop of j_c with decreasing JJ size can be observed, even if the physical junction size is measured using an SEM. This behavior is depicted in Fig. 3.12 as black circles. The colored symbols represent calculated j_c-values, taking different linear shrinkages into account. This linear correction factor affects junctions of smaller sizes much more strongly than those with large areas, leading to a strong deviation towards very small JJs. Fig. 3.12 shows j_c-values for $s_{JJ} = 250\,nm$ in green, 320 nm in red and 450 nm in blue symbols. Again, a shrinkage of $s_{JJ} \approx 300\,nm$, yields an almost constant critical-current density over all junc-

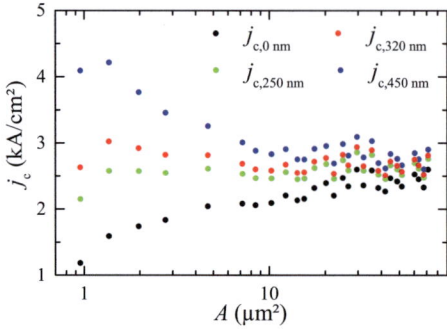

Figure 3.12.: Critical-current densities extracted from JJs of various sizes (black circles). The colored symbols correspond to different values for linear shrinkage correction.

tion sizes. Despite the stronger effect on small JJs, there is still a significant impact even on large JJs, leading to an increase of the calculated critical-current density from approximately $2.1 \, \text{kA/cm}^2$ to roughly $2.5 \, \text{kA/cm}^2$. This fits the results from the linear fit if the square root of the critical current over the square root of the junction size, discussed earlier. In conclusion, it can be inferred that the Josephson junctions fabricated exhibit a good homogeneity and reproducibility over the length scale of several mm and are thus suitable for a large variety of Josephson devices incorporating short JJs.

The inductance of the wiring layer can be extracted by interferometer measurements [67, 68, 113]. Essentially, the device consists of a SQUID, which is modulated by a magnetic field Φ_{cl} created by a control-line current I_{cl} passing in the wiring layer, depicted in Fig. 3.13 (a). The inductance may then be extracted from one period of the $I_c(\Phi_{cl})$ dependence:

$$\Delta I_{cl} = \frac{\Phi_0}{L_1 + L_2}. \tag{3.3}$$

Here $L_{1,2}$ are the inductances in the wiring layer, c.f. Fig. 3.13 (b). It should be noted that Eq. (3.3) holds only under the assumption that the inductances of the lower electrode $L_{3,4} \ll L_{1,2}$, which may be achieved by placing the entire structure on a bottom electrode, which is significantly wider than the control line. Additionally, the London penetration depth for the lower electrode and wiring layer may be extracted from such measurements using

$$L_1 + L_2 = \mu_0 \cdot \frac{L_w}{W_w} \left[t_{ins} + \lambda_L \coth\left(\frac{d_b}{\lambda_L}\right) + \lambda_L \coth\left(\frac{d_w}{\lambda_L}\right) \right], \tag{3.4}$$

where μ_0 is the vacuum permeability, L_w is the length and W_w the width of the inductance in the wiring layer.

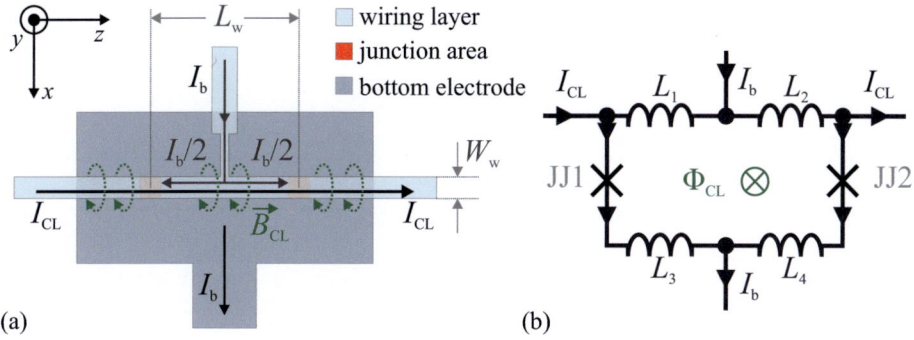

Figure 3.13.: Interferometer designs and measurements for the extraction of the mutual inductance and London penetration depth. (a) shows a schematic top view of the interferometer device, while in (b) the equivalent circuit diagram is given.

3.3.3. Characterization of Long Josephson Junctions

In the previous section the high quality of short junctions has been discussed. However, as described in section 2.4, most of the devices of interest in this thesis are based not only on short JJs, but also on long junctions. Before the beginning of this work the question, whether the barrier in a LJJ shows inhomogeneities, which may cause premature switching, remained unanswered. In order to probe the barrier quality and homogeneity on a length scale of λ_J and above, a new experiment had to be derived. To achieve maximum precision, the exact geometry of the fabricated junctions and the resulting change for the fluxon dynamics, was taken into account [135]. All of the junctions discussed in the following were fabricated in the overlap geometry, where there is an overlapping region of the wiring electrode and the ground electrode surrounding the active JJ area. These regions are called idle regions and affect the Josephson penetration depth according to:

$$\lambda_{J,\text{eff}} = \lambda_J \sqrt{1 + \frac{2w_2}{w_1} \cdot \frac{t_{\text{ox,eff}}}{t_{\text{ins,eff}}}}, \tag{3.5}$$

where $w_{2,1}$ correspond to the width of the idle region and the active region (*i.e.* junction width), respectively [136]. $t_{\text{ins,eff}}$ and $t_{\text{ox,eff}}$ are the effective thicknesses of the idle and active regions, calculated by:

$$\begin{aligned}
t_{\text{ox,eff}} &= t_{\text{ox}} + \lambda_L \coth\left(\frac{d_b}{\lambda_L}\right) + \lambda_L \coth\left(\frac{d_t + d_w}{\lambda_L}\right), \\
t_{\text{ins,eff}} &= t_{\text{ins}} + \lambda_L \coth\left(\frac{d_b}{\lambda_L}\right) + \lambda_L \coth\left(\frac{d_w}{\lambda_L}\right),
\end{aligned} \tag{3.6}$$

where d_b, d_t and d_w denote the thicknesses of the bottom electrode, of the top electrode of the trilayer and the thickness of the wiring layer respectively. t_{ins} is the total thickness

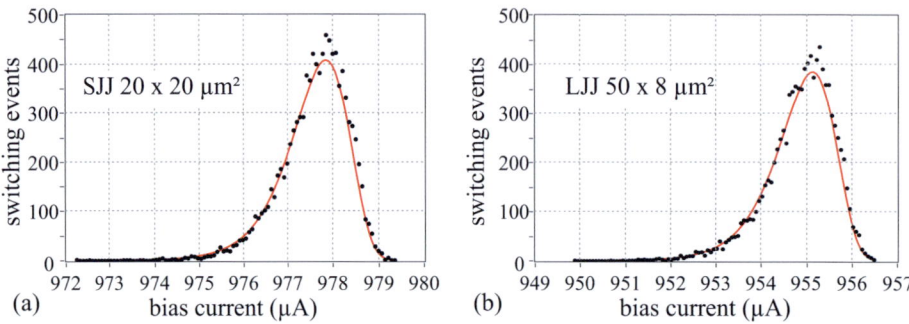

Figure 3.14.: Histograms of 10000 switching events for JJs of comparable area but different geometry. (a) Depicts the histogram of a short JJ, whereas (b) shows the histogram of a long JJ. The small difference in the standard deviations ($\sigma_{20\times20} = 0.750\,\mu A$ and $\sigma_{50\times8} = 0.749\,\mu A$), can be explained by a small difference in the effective JJ size.

of insulation layers separating the bottom electrode from the wiring layer as indicated in Fig. 3.4. The effective thicknesses calculated by Eq. (3.6) assume that all metallizations are of the same superconductor, exhibiting the identical London penetration depth λ_L in all layers. Typical values for the London penetration depth range from 80 to 90 nm [111–113]. For most calculations in this thesis, λ_L has been taken to be 90 nm.

Before the investigation of very long junctions, first, junctions of intermediate length at the order of λ_J, were investigated in comparison to their short counterpart of identical area. For this, a design including several pairs of short and long junctions of identical junction area was developed, and the switching characteristics were investigated. For short JJs with all dimensions shorter than λ_J, the Josephson phase can be considered point-like and thus will not be influenced by barrier inhomogeneities on the length scale of λ_J. However, junctions of intermediate length will be affected. A comparison of switching histograms of such a pair of JJs will therefore contain information on the *long range* homogeneity of the tunneling barrier. In Fig 3.14, histograms of 10000 switching events measured at 4.2 K are shown for a $20 \times 20\,\mu m^2$ and $50 \times 8\,\mu m^2$ JJ are shown. At a critical-current density of $j_c = 260\,A/cm^2$ and a resulting Josephson penetration depth of $\lambda_J \approx 25\,\mu m$ (depending on the exact geometry), the first corresponds to a short JJ, whereas the second is already at the verge of the long junction regime. In case of a homogeneous barrier, σ should be equal for both JJs, and follow the dependence given in Eq. (2.27). The slight difference between $\sigma_{20\times20} = 0.750\,\mu A$ and $\sigma_{50\times8} = 0.749\,\mu A$ is due to the slight difference of the effective junction area, which has been measured using SEM.

In Fig. 3.15 the standard deviations of three pairs of junctions are plotted. Again, each pair consists of a $20 \times 20\,\mu m^2$ (depicted as solid circles) and $50 \times 8\,\mu m^2$ (depicted as open circles) JJ. Different normalized lengths are achieved by a variation of the critical-current densities $j_{c,1} \approx 78\,A/cm^2$, $j_{c,2} \approx 163\,A/cm^2$ and $j_{c,3} \approx 260\,A/cm^2$, extracted from the IVC and the measured JJ-area. The dotted gray line indicates the theoretical proportionality

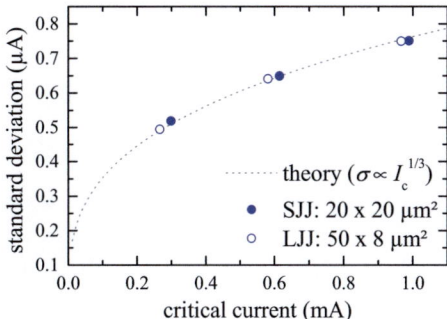

Figure 3.15.: Standard deviation of 10000 switching events measured for three pairs of $20 \times 20\,\mu m^2$ (depicted as solid circles) and $50 \times 8\,\mu m^2$ (depicted as open circles) JJs. The dotted gray line shows the theoretical proportionality of $\sigma \propto I_c^{1/3}$.

$\sigma \propto I_c^{1/3}$. The deviation from theory is within the measurement accuracy. From this it can be concluded, that the barrier is homogeneous not only for short junctions with $W = L \ll \lambda_J$, but also on a length scale of the Josephson penetration depth.

For junctions with $\lambda_J \gg 1$, the analysis based on switching histograms was not possible, since the technological process at that stage of this work did not allow reliable fabrication of such narrow junctions. Instead, a new biasing scheme for LJJs has been developed, which allows a comparison to the classical approach using a wide superconducting line feeding the bias current to the LJJ. Due to the current profile in a wide SC biasline, which can be approximated by

$$\iota_{\tilde{z}} = \frac{\iota_0 l}{\pi \sqrt{\tilde{z}(l-\tilde{z})}}, \tag{3.7}$$

where ι_0 is the mean linear current density in units of $j_c \cdot W$ [137, 138], the effective critical current of a long JJ is significantly suppressed as compared to the theoretical critical current calculated from j_c and the JJ-area A. Consequently, the normalized critical current $i_n = I_c/\lambda_J W j_c$ scales proportional to $(L/\lambda_J)^{1/2}$. The measured dependence $i_n(l)$ for classical bias lines is plotted as black circles in Fig. 3.16 (a). The theoretical curve $i_{n,theory} = 2.35(l)^{0.5}$ is indicated as a dotted gray line. For the fit, three measurement points (indicated in dark green) have been neglected due to a significant amount of trapped flux during the measurement. The resulting fit is plotted as a dashed black line and follows the dependence $i_{n,fit} = 2.31(l)^{0.49}$, which is reasonably close to theory.

The new biasing scheme developed within this thesis is based on a resistive circuitry in series to the LJJ as shown schematically in Fig. 3.16 (b). Instead of a wide superconducting biasline, multiple superconducting wires are connected in parallel between the junction and a resistor. Each of these lines carries the same current since the bias current profile within the resistor material is constant along the coordinate z. By this the inhomogeneous current profile from Eq. (3.7), schematically indicated in the wide biaslines by arrows of varying

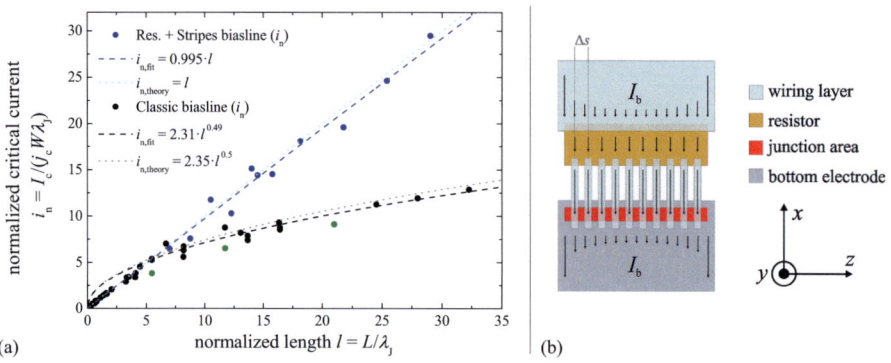

Figure 3.16.: (a) Effects of homogeneous bias current injection using striped feed line and serial resistor. In the long junction regime the normalized critical current follows the proportionality $i_n = I_c/\lambda_J W j_c \propto (L/\lambda_J)^{1/2}$ for classical wide bias lines, whereas the newly developed bias circuitry, depicted in (b), results in a direct proportionality of $i_n \propto L/\lambda_J$ for short and long Josephson junctions.

length, is compensated. Fig. 3.16 (a) shows the measured critical currents of LJJs based on such bias lines as blue circles. The blue dashed line corresponds to the linear fit of the data, given by $i_n = 0.995 \cdot l$. For comparison $i_n = l$ is also plotted as a dotted blue line. The slightly smaller critical currents in both cases (classical bias and resistive bias) can be explained by trapped flux in almost all measurements, since junctions of these lengths are extremely susceptible to stray magnetic fields. Furthermore, the redistribution of the bias current towards the edges of the bottom electrode causes an additional suppression of I_c even for the homogenized current feeds.

The key dimensions for the new biasing are dominated by λ_L of the top JJ electrode and λ_J. Within a superconductor the current redistributes to the edges on a length scale of λ_L, which makes it necessary to end the finger on the top electrode of the junction which is at the order of λ_L. However, the length scale on which the current redistributes within the junction is given by λ_J. Although for dc characterization of a LJJ this makes only one finger per λ_J necessary, it is advisable to fabricate a more closely spaced design. First, this introduced a redundancy compensating for lithographic errors and furthermore, even contracted fluxons traveling at high speeds still experience an almost homogeneous Lorentz force due to this homogenized current profile. Thus, for closely spaced fingers ($\Delta s \ll \lambda_J$) placed on the junction the idealized biasing presented in section 2.3 Fig. 2.9 (b) is simulated. The crossover to LJJ-regime can be approximated by the intersection of the two fits depicted as dashed lines at $l \approx 5.3$. The theoretical curves are depicted as dotted lines.

The occurrence of Fiske steps in LJJs has been discussed theoretically in section 2.3. In Fig. 3.17 a typical measurement of such steps is depicted. Using Eq. (2.50) and the position of the nth-step along the voltage-axis, the specific capacitance of a given trilayer

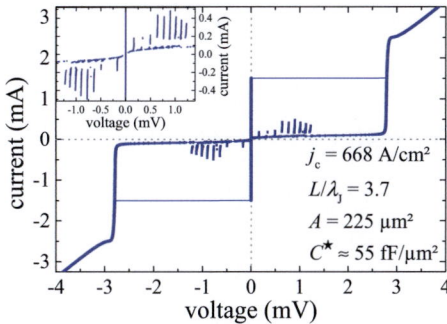

Figure 3.17.: Fiske steps measured on a LJJ with $l = 3.7$. From the position of the Fiske steps the specific capacitance of the trilayer can be calculated.

may be extracted. For the shown example of a trilayer with $j_c = 668\,\mathrm{A/cm^2}$, the specific capacitance has been found to be $C^\star \approx 55\,\mathrm{fF/\mu m^2}$.

Conclusion

Using short and long Josephson junctions, it is possible to extract important parameters, characteristic for a particular fabrication process. One of the most important parameters is the already mentioned critical-current density. It can be approximated from IVC measurements $j_c = I_c/A$. Alternatively, the measurements of serial arrays of JJs or fitting of various JJs with different sizes offers a more precise extraction. By the latter the linear shrinkage may also be extracted and accounted for.

The measurement of switching histograms allowed a clear comparison of the switching dynamics of short and intermediate-length junctions. Variations in the effective junction sizes were taken into account by previous measurements of the JJ-area using SEM. From the switching histograms the standard deviation has been extracted for three junction pairs with different critical-current densities consisting of a $20 \times 20\,\mathrm{\mu m^2}$ and $50 \times 8\,\mathrm{\mu m^2}$ JJs. The obtained dependence of the standard deviation of the histograms follows the theoretical behavior precisely. Consequently, it can be concluded, that junctions at the length scale of λ_J do not suffer from premature switching, as compared to short JJs, due to possible inhomogeneities in the tunneling barrier.

With the given technology at that point of this thesis, it was not possible to conduct such investigations for significantly longer junctions. To enable investigation of very long JJs $L \gg \lambda_J$, a new biasing scheme, based on a resistive circuitry, has been developed. Using this, it is possible to fabricate LJJs with an almost ideally homogeneous bias current profile. This results in an enhanced critical current as compared to a classical biasing circuitry using a wide superconducting bias line. The extracted dependences for both the classical, as well as the newly developed bias circuitry, fit to theory. This indicates that the tunneling barrier is almost free of defects, even on a length scale significantly exceeding the Josephson penetration depth.

3.4. Investigated Josephson Junction Devices

Next, the devices based on Nb/Al-AlO$_x$/Nb trilayers developed and investigated throughout this thesis will be described in detail and the results obtained will be discussed. The first subsection deals with the work done on dc SQUIDs with a strong R-asymmetry that were introduced in subsection 2.4.1. The second part is dedicated to different fractional vortex devices described theoretically in subsection 2.4.2. The chapter ends with the discussion of the obtained results from the investigation of flux-flow oscillators as described in subsection 2.4.3. In particular the effects of an LC-shunt to a FFO and its influences on the emission linewidth will be discussed.

3.4.1. Enhanced Superconducting Quantum Interference Devices

In subsection 2.4.1 the SQUID was introduced as a powerful device to measure extremely small magnetic fields. At the end of that section the possibility to introduce asymmetries in various design parameters was mentioned as a mean of steepening the transfer function of the device, which results in a more pronounced voltage response with respect to changes in the external magnetic flux. Within this work, symmetric and asymmetric overdamped dc SQUIDs have been fabricated and investigated [RNM$^+$12].

Introducing asymmetries as described in Eq. (2.57) to the Langevin equations (Eq. (2.54)), results in

$$\frac{i}{2} + i_{\text{circ}} = (1 - \alpha_{\text{I}})\sin\varphi_{\text{JJ1}} + (1 - \alpha_{\text{R}})\dot{\varphi}_{\text{JJ1}} + \beta_{\text{c}}(1 - \alpha_{\text{C}})\ddot{\varphi}_{\text{JJ1}} + \gamma_{\text{N,JJ1}}$$
$$\frac{i}{2} - i_{\text{circ}} = (1 + \alpha_{\text{I}})\sin\varphi_{\text{JJ2}} + (1 + \alpha_{\text{R}})\dot{\varphi}_{\text{JJ2}} + \beta_{\text{c}}(1 + \alpha_{\text{C}})\ddot{\varphi}_{\text{JJ2}} + \gamma_{\text{N,JJ2}},$$

$$(3.8)$$

and

$$\varphi_{\text{JJ2}} - \varphi_{\text{JJ1}} = 2\pi\Phi_{\text{a}} + \pi\beta_{\text{L}}\left(i_{\text{circ}} - \frac{\alpha_{\text{L}}}{2}\gamma\right). \tag{3.9}$$

In case of a sub-critical bias current ($I < I_{\text{c}}$), all time dependent parameters may be neglected. From this it may thus be inferred, that α_{R} and α_{C} do not affect the $I_{\text{c}}(\Phi)$ dependence. However, asymmetries in I and L cause a shift of the $I_{\text{c}}(\Phi_{\text{a}})$ dependence along the Φ-axis [139, 140]. From measurements of $I_{\text{c}}(\Phi_{\text{a}})$ for positive and negative bias, the sum of these asymmetries can be calculated by $\alpha_{\text{I}} + \alpha_{\text{L}} = \Delta\Phi/\Phi_0\beta_{\text{L}}$, where $\Delta\Phi$ denotes the shift between the positive and negative curve [14]. This shift is caused by the effective flux generated by either an increased current or inductance in one of the two SQUID arms which couples to the SQUID even for $\Phi_{\text{a}}/\Phi_0 = 0$.

Table 3.2.: Summary of the key parameters of the SQUID amplifier used of extraction of the noise performance of the asymmetric SQUIDs.

Parameter	Value
critical current ($2I_0$)	$77\,\mu A$
total resistance (R)	$0.7\,\Omega$
screening parameter (β_L)	2.3
damping parameter (β_c)	0.5
flux noise ($S_\Phi^{1/2}(f)\vert_{f>10\,\mathrm{Hz}}$)	$1.5\,\mu\Phi_0/\sqrt{\mathrm{Hz}}$
input coil inductance (L_i)	$105\,\mathrm{nH}$
inductive coupling ($1/M_i$)	$0.93\,\mu A/\Phi_0$

In order to extract α_L by itself, an over-critical bias measurement is necessary. A shift of the $V(\Phi)$ characteristics is proportional to $1/2 \cdot \alpha_L LI$ [14]. It should be noted, that I-asymmetries and C-asymmetries are closely related, since any change in the junction size equally affects the critical current and the junction capacitance ($I_c \propto A$ and $I_c \propto C$). Certainly, the possibility of a capacitive shunt [47] or two separately deposited trilayers allow separate tuning of these parameters, however this has not been done within this thesis. Consequently, for all 2-JJ SQUIDs discussed hereafter, $\alpha_I = \alpha_C$ may be assumed.

All the measurements shown in the remaining part of subsection 3.4.1 were performed at the *Institut für Experimentalphysik II* (PIT II) at the *University of Tübingen*, Germany. The setup is also based on a dipstick design, including a copper shielding for suppression of rf interferences and a cryoperm shielding for suppression of stray magnetic fields. This results in a remanent magnetic field of $B_{rem} \approx 100\,\mathrm{nT}$ at the sample [139]. For modulation of the SQUID, the sample holder is equipped with a superconducting coil underneath the sample, such that an applied dc current causes a magnetic field normal to the SQUID-loop up to $B_{a,max} \approx 1\,\mathrm{mT}$ [140]. The dc characterization measurements IVC, $I_c(\Phi_a)$ and $V(\Phi_a)$ have been performed using a room temperature voltage amplifier (RTA). Applied currents for the bias, the superconducting coil and the read-out voltage, have been controlled with the software *GoldExI* [132]. For characterization of the noise performance, the RTA was replaced by a low-temperature SQUID amplifier (SA), which is part of a commercial SQUID amplifier system [141]. Parameters of the SQUID amplifier are given in Tab. 3.2. In order to achieve optimal performance the device under test (DUT), including the LHe dewar was placed inside an rf shielding room during noise measurements. A more detailed description of the setup is given in [139, 140].

Three different measurement modes are possible using the SA. The most basic is called the open loop (OL) mode, in which the SA is biased at the steepest point of the amplifier

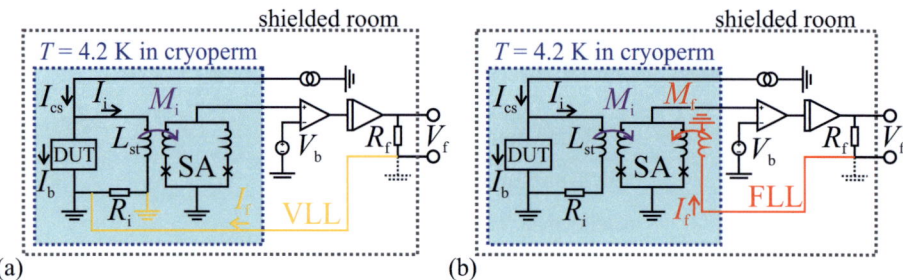

Figure 3.18.: Different measurement modes of a SQUID amplifier. (a) Depicts the voltage-locked loop mode, for which the flux in the amplifier SQUID is kept constant by compensating the input current I_i to zero using the feedback current I_f. (b) Shows the flux-locked loop mode, where the amplifier SQUID is kept at the optimal point of V_Φ.

SQUID's transfer function, i.e. $|\partial V_{SA}/\partial \Phi_a| = \max$. For a small swing of Φ_a the function may be considered to be linear, leading to the approximate proportionality $V_{SA} \propto \Phi_a$. The limit of this mode is clearly given by the range for which the transfer function may be linearized. For open loop mode the gain is given by $G_{OL} = V_\Phi M_i/R_i$ and is typically in the range of 100. Here M_i denotes the mutual inductance of the input coil to the amplifier SQUID and R_i is the serial resistance in the input coil. The open loop mode is depicted in black in Fig. 3.18.

The second measurement mode, called the voltage-locked loop mode (VLL), is shown as a yellow expansion in Fig. 3.18 (a) to the OL mode. In this mode the flux in the amplifier SQUID is kept constant, by applying a feedback current I_f to the input coil, such that the total current through the input inductance L_i is zero. Consequently, the voltage across the SQUID being characterized is fixed. This voltage-locked loop mode brings the advantage of a high input impedance due to $I_i = 0\,\text{mA}$, leading to a large voltage gain $G_{VLL} = |-R_f/R_i|$ typically in the range of $G_{VLL} = 10^4$ [140]. However, VLL comes at the cost of a galvanic coupling of the device to the read-out electronics, often introducing grounding problems and consequently significant additional noise. The most crucial disadvantage of the VLL mode is however, the dependence of the amplifier SQUID noise $S_{V,SA}^{1/2}$ on the resistance of the device under test. The voltage noise of the SA in VLL mode is given by:

$$S_{V,SA,VLL}^{1/2} = \left[4k_B T R_i + \left(\frac{R_i + R_{DUT}}{M_i} \right)^2 \cdot S_\Phi \right]^{1/2}. \tag{3.10}$$

Characterization of devices with comparable voltage noise to that of the SA thus becomes very complicated since the noise from the SA itself is difficult to subtract from the measurements, due to the dependence of $S_{V,SA}^{1/2}$ on R_{DUT}.

For very low noise devices, such as the SQUIDs investigated within this thesis, the third measurement mode, the so-called flux-locked loop (FLL) mode is preferred, shown in Fig. 3.18 (b) as a red expansion to the OL mode. Unlike in the VLL, the voltage noise of the SA is given by:

$$S_{V,SA,FLL}^{1/2} = \left[4k_B T R_i + \left(\frac{R_i}{M_i} \right)^2 \cdot S_\Phi \right]^{1/2} \tag{3.11}$$

and is therefore independent of the resistance of the DUT. As a result, the measurements can be corrected easily, by subtracting the constant noise parameter of the SA. An additional advantage of the FLL mode is the very high gain values determined by $G_{FLL} = R_f M_i / R_i M_f$. For typical values of $R_f \approx 10\,k\Omega$, $R_i \approx 1\,\Omega$, $M_i \approx 1\,\Phi_0/\mu A$ and $M_f \approx 10\,\Phi_0/\mu A$ the gain is in the range of $G_{FLL} = 10^5$ [139, 140]. Usually, the main drawback of the FLL mode is considered to be the low input impedance, which is primarily determined by $R_i^{-1} + R_{DUT}^{-1}$. To compensate this, it is necessary to counterbalance the current I_i artificially, by adjusting the current I_{cs}, which splits into $I_{cs} = I_b + I_i$. From measurements of V_f, the input current can be extracted according to $I_i = M_f V_f / M_i R_f$, making it necessary to precisely extract the coupling ratio M_f/M_i, which varies slightly for each DUT. Since I_i depends on the bias current I_b and the applied flux Φ_a, the current I_{cs} has been iterated such that $I_b = I_{cs} - I_i = $ const. is true for all Φ_a. For details on the measurement procedure and the extraction of the coupling ratio M_f/M_i, it is referred to [139, 140][RNM$^+$12]. It should be noted, that this effect of a non-constant input current exists also for symmetric SQUIDs, it is however much less pronounced as compared to its asymmetric counterpart, due to the symmetric current-flow in the two SQUID arms. In case of an asymmetric device, the predominant current flow in one of the arms and the resulting additional flux coupled to the SQUID loop, causes a significant change in the $V(\Phi_a)$ characteristics. Using this novel biasing scheme, it is possible to measure even strongly asymmetric dc SQUIDs in FLL mode without the drawback of a low input impedance.

α_R was the most investigated asymmetry parameter within this thesis, although SQUIDs with various other asymmetries were fabricated as well. The asymmetry in the resistance causes an asymmetric $V(\Phi_a)$ characteristic, making one of the slopes much steeper than the other and thus causing an increase in the transfer function V_Φ. In order to achieve an improved performance of the SQUID, it is however necessary to take the effects of noise into account. The value of interest is the normalized energy resolution e_r of the SQUID. This value takes into account all parameters affecting the overall performance with respect to noise influences. The Nyquist noise current in the shunt resistors is given by:

$$S_I(f) = \frac{4k_B T}{R}, \tag{3.12}$$

where R is the total resistance of DUT. In order to ensure good Josephson coupling, the condition $I_0 \Phi_0 / 2\pi \gg k_B T$ needs to be fulfilled, *i.e.* the Josephson coupling E_J should be significantly larger than the energy of the thermal noise E_{th}. This ratio is also called the

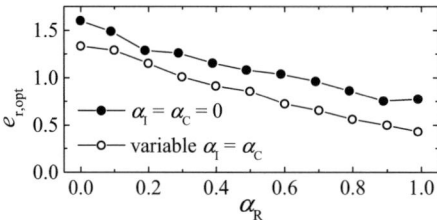

Figure 3.19.: Numerical simulations of asymmetric Langevin equations Eq. (3.8) with respect to R-asymmetries. The solid black circles depict the case of $\alpha_I = \alpha_C = 0$, whereas the open circles show the normalized energy resolution for individually optimized $\alpha_I = \alpha_C$. Fixed parameters are $\Gamma_N = 0.01$ and $\beta_c = 0.7$. β_L has been varied for each point [RNM+12].

noise ratio

$$\Gamma_N = \frac{2\pi k_B T}{I_0 \Phi_0} = \frac{I_N}{I_0} \tag{3.13}$$

and should be kept well below 1, for pronounced Josephson effects. The noise current introduces a voltage noise S_V across the SQUID as well as a spectral density of equivalent flux noise S_Φ coupled to the loop, which is given by:

$$S_\Phi = \frac{S_V}{|V_\Phi|^2}. \tag{3.14}$$

The flux noise poses the figure of merit for the device performance. Phenomenologically speaking it represents the trade-off between the inherent noise and the transfer function, thus determining the resolution. From this, the energy resolution can be calculated to $\varepsilon = S_\Phi/2L_{sq}$, where L_{sq} is the inductance of the SQUID [139]. For an optimized device this value should be as small as possible in order to achieve high sensitivity. For devices with $\beta_c = \beta_L = 1$ the energy resolution is in the range of $\varepsilon \approx 9 \cdot (k_B T L_{sq}/R)$ [14][RNM+12]. In normalized units the energy resolution may be re-written to:

$$e_r = \frac{s_\phi}{2\Gamma_N \beta_L}, \tag{3.15}$$

where $s_\phi = S_\Phi I_0 R / 2\Phi_0 k_B T$, Γ_N is the noise ratio and β_L is the screening parameter.

Fig. 3.19 depicts numerical simulations of the Langevin equations Eq. (3.8) performed at the PIT II, *University of Tübingen*. As can be seen, the effects of R-asymmetries promise a significant decrease in e from approximately 1.6 to 0.7. Thus the resulting device with $\alpha_R \to 1$ should have an energy resolution almost twice as good as compared to the symmetric version ($\alpha_R = 0$). Although the effects of current and capacitance asymmetries seem to be quite desirable, it turns out that for practical reasons theses asymmetries will most likely not show such pronounced effects as predicted in simulations [140][RNM+12]. Fabrication of devices exhibiting a large α_R is on the other hand rather simple from the design point of

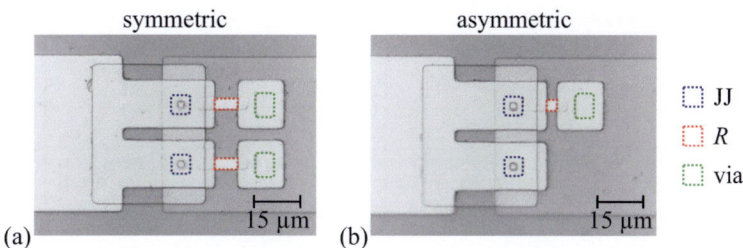

Figure 3.20.: Micrographs of a symmetric and an asymmetric dc SQUID in (a) and (b), respectively.

view. Whereas the conventional SQUID consists of two Josephson junctions individually shunted resulting in R_1 and R_2 such that $\beta_c \leq 1$ is maintained, an asymmetric SQUID may be realized by simply removing the shunt from JJ_1 entirely and replacing the shunt on JJ_2 by a shunt resistor yielding $R^{-1} = R_1^{-1} + R_2^{-1}$. The asymmetry is finally only limited by the inherent junction resistance of the unshunted junction. For high-quality tunneling barriers and reasonably small JJ, the ratio between R_1 and R_2 can reach values in the range of $R_1/R_2 \approx 10000$ [140].

From the point of view of fabrication, the goal has been to design SQUIDs with shunt resistors exhibiting a minimal parasitic inductance L_p. Additionally, any remaining inductance should be affected by the applied magnetic flux as little as possible. These requirements were inferred from preceding measurements of dc SQUIDs at the PIT II, for which the parasitic inductance was comparable to the SQUID inductance and was strongly affected by Φ_a [139]. In most JJ fabrication processes currently commercially available, shunt resistors are placed next to the junction electrodes according to individual design rules of the corresponding foundry. Mostly, this is done in order to reduce parasitic capacities in parallel to the JJs. Nevertheless, this results in rather long interconnects between the junction and R_{shunt}, introducing a parasitic inductance L_p at the order of L_{sq}. Furthermore, this parallel arrangement forms a loop adjacent to the JJ, representing an additional LC resonance circuit with the SQUID. Consequently, the SQUID characteristics are superposed with the LC characteristics in case of comparable frequencies $f_J = V/\Phi_0$ and $f_{LC} = 1/2\pi \cdot \sqrt{LC/2}$. For $\Phi_a = 1/2 \cdot \Phi_0$, the resonance effect gets strongest since the two JJs oscillate in anti-phase [139, 142].

In order to avoid such resonance effects, the SQUID design required a negligible L_p, shielded from externally applied magnetic fields. Fig. 3.20 shows micrographs of a symmetric and asymmetric SQUID in figure (a) and (b), respectively. The shunt resistors have been placed directly in series to the junctions, only a few μm from the JJs and the vias and thus minimizing the inductance. For minimal effects of Φ_a on the shunt, the entire structure has been placed on top of the ground electrode. The Meissner Ochsenfeld effect effectively shields the externally applied magnetic field from the shunt loop. As compared to the above described parallel arrangement, the parasitic inductance here can be assumed to be zero.

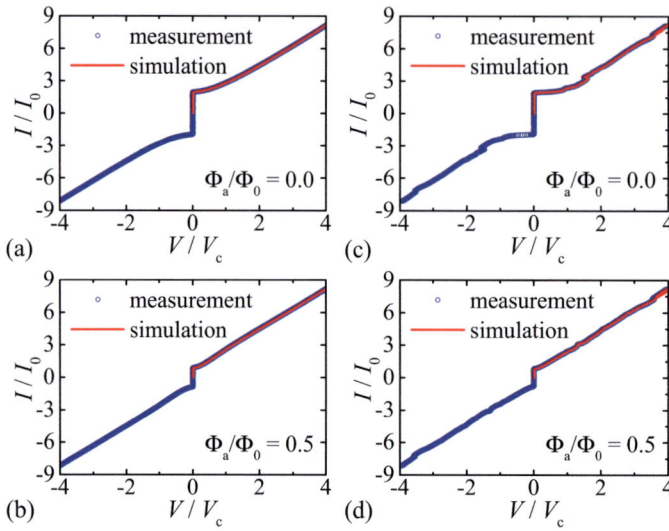

Figure 3.21.: *IVC* measurements and numerical calculations of (a) symmetric SQUID for $\Phi_a/\Phi_0 = 0.0$ and (b) $\Phi_a/\Phi_0 = 0.5$. Figure (c) shows the asymmetric counterpart for $\Phi_a/\Phi_0 = 0.0$ and (d) for $\Phi_a/\Phi_0 = 0.5$. The open blue symbols correspond to measurement data and the red line to numerical simulations performed at the PIT II.

As a result the characteristics of the SQUIDs can be calculated numerically, based on the asymmetric Langevin equations introduced in Eq. (3.8).

Fig. 3.21 shows *IVC* measurements and numerical calculations of a symmetric (sample S) and an asymmetric (sample A) SQUID. The parameters of the two devices are summarized in Tab. 3.3 [139][RNM+12]. The lack of any anomalies in the *IVC*s of the symmetric device (a) and (b), indicates that the effects of parasitic inductances due to the shunt structure are minimal even for an applied field of $\Phi_a/\Phi_0 = 0.5$. The *IVC*s of the asymmetric sample (b) and (c) show the peculiarities expected from such devices. The region with negative differential resistance at approximately $3.4I_0$ in (c) marks the crossover from a stable low

Table 3.3.: Parameters of the symmetric SQUID (sample S) and the asymmetric SQUID (sample A).

Parameter	sample S	sample A
critical current ($2I_0$)	64.4 µA	62.4 µA
total resistance (R)	0.54 Ω	0.57 Ω
characteristic voltage (V_c)	37.07 µV	35.3 µV
screening parameter (β_L)	0.74	0.675
inductance (L_{sq})	23.78 pH	22.4 pH

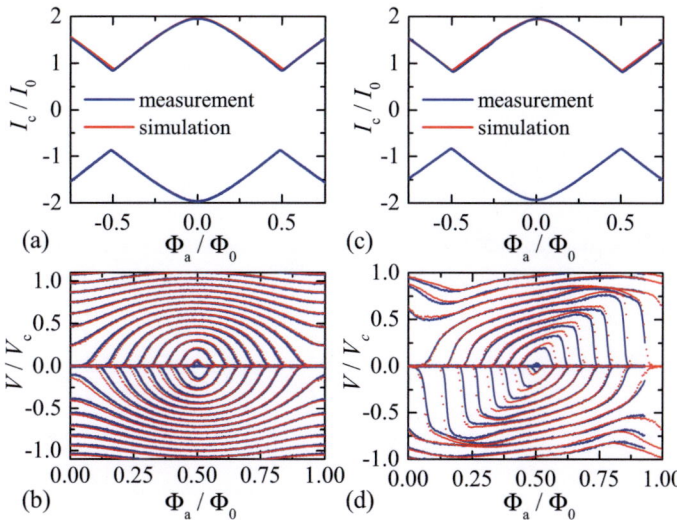

Figure 3.22.: $I_c(\Phi_a)$ measurements of symmetric (a) and asymmetric (c) dc SQUIDs. In (b) and (d) the $V(\Phi_a)$ characteristics are displayed for the symmetric and asymmetric sample, respectively. The blue traces correspond to measurement data recorded using a RTA and the red trace / symbols to numerical simulations based on the Langevin equations.

bias current regime to a high bias current regime exhibiting chaotic dynamics [RNM+12]. As desired, there are no additional anomalies due to parasitic LC-shunts, making it possible to simulate the behavior using equations 3.8, assuming a $\beta_{L,S} = 0.740$ and $\beta_{L,A} = 0.675$. From this the inductance can be extracted to be $L_A = 22.4\,\mathrm{pH}$, fitting the design value of $L_{des} = 23.90\,\mathrm{pH}$ reasonably well. As some deviation of j_c from the design parameter was anticipated, the design of the $10 \times 10\,\mathrm{mm}^2$ chip included a wide parameter spread over 29 different pairs of symmetric and asymmetric devices. Therefore, merely one fabrication run was necessary for working devices with the required parameter set.

From the measurements of the $I_c(\Phi_a)$ characteristics, the sum of the current- and inductance-asymmetry can be extracted. Fig. 3.22 (a) and (c) show the corresponding curves for sample S and A normalized to Φ_0. The total shift along the Φ_a/Φ_0-axis between the positive and negative absolute maxima is approximately $\Delta\Phi \approx 0.013\Phi_0$. Based on the screening parameters extracted from simulations, c.f. Tab. 3.3, this corresponds to a combined $L - I$-asymmetry of $\alpha_I + \alpha_L < 2\%$. Further details of the simulation procedure and parameters are given in [139, 140][RNM+12].

Fig. 3.22 (b) and (d) show the corresponding voltage modulations with respect to Φ_a. The symmetrically shunted SQUID exhibits a symmetric $V(\Phi_a)$ characteristic, with equally moderate slopes on both sides. In striking contrast, the asymmetrically shunted SQUID exhibits a $V(\Phi_a)$ characteristic with one slope significantly steeper than the other. The behavior of both samples was measured using the RTA and can be simulated reasonable well based on Eq. (3.8).

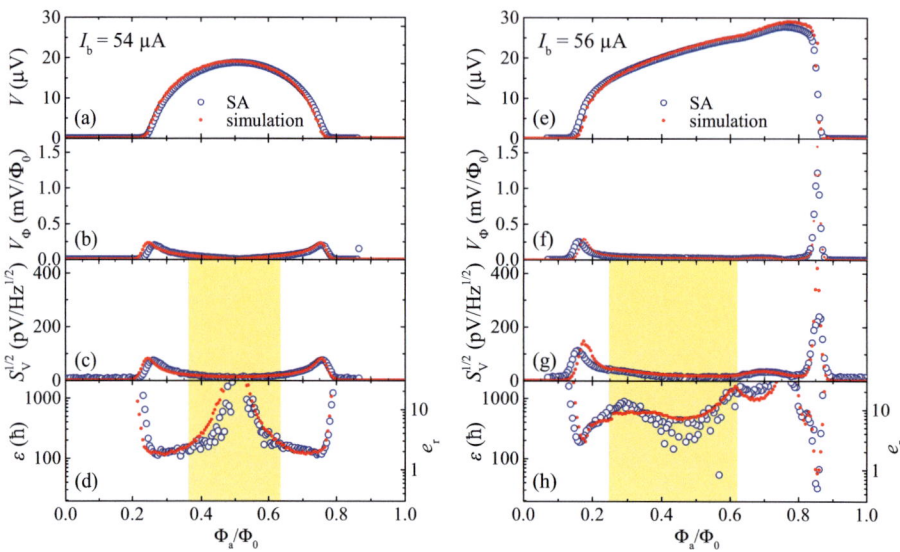

Figure 3.23.: (a) And (e) show the $V(\Phi_a)$, (b) and (f) the $V_\Phi(\Phi_a)$, (c) and (g) the $S_V^{1/2}(\Phi_a)$ and (d) and (h) the $\varepsilon(\Phi_a)$ as well as the $e_r(\Phi_a)$ dependences for a symmetric and asymmetric SQUID, respectively. The shaded area corresponds to the measurement regime where the noise level of the SQUID amplifier is equal or even larger than that of the DUT. Extracted values for ε and e_r are consequently not reliable in this region. The lowest value of e_r for the symmetric SQUID is $e_{r,S} = 1.79$, which is approximately 3.4 times larger than that of the novel asymmetric SQUID exhibiting a normalized energy resolution of $e_{r,A} = 0.52$.

For the concluding measurements of the noise performances of the two SQUIDs, the energy resolutions were extracted from V_Φ and white voltage noise $S_V^{1/2}$ measurements. $S_V^{1/2}$ was averaged in a frequency band from 100 Hz to 3 kHz for various different bias currents from $40 - 70$ mA (in case of the asymmetric SQUID only up to $I_b = 68$ mA). Fig. 3.23 (a) and (e) show the $V(\Phi_a)$ characteristics for the optimal bias currents $I_{b,S} = 54\,\mu$A and $I_{b,A} = 56\,\mu$A, respectively. In addition to the measurement based on the novel FLL biasing scheme described above (blue trace), the numerical calculations based on Eq. (3.8) (red trace) are shown. The traces coincide well, indicating that the FLL mode with an extra correction current to simulate a high input impedance, $i.e.$ $I_i = 0$ mA, meets the requirements for high precision measurements. Figures (b) and (f) depict the transfer functions $V_\Phi = \partial V/\partial\Phi_a$. It should be noted that the maximum value for the asymmetric device reaches up to $V_\Phi = 1.22$ mV, which is comparable to SQUIDs fully optimized for amplifier devices, $i.e.$ to reach the highest possible V_Φ [143]. Figures (c) and (g) show the white voltage noise. The measurement was performed in the FLL mode with an extra applied correction current, and the constant noise of the SQUID amplifier was subtracted. From these measurements the absolute energy resolution ε, as well as the normalized energy resolution e_r, was calculated based on Eqs. (3.15), (3.13) and (2.55). Both dependences are plotted over the applied magnetic flux in Fig. 3.23 (d) and (h) for the symmetric and asymmetric device, respectively. The symmetric SQUID exhibits an energy resolution of $\varepsilon = 110\hbar$, corresponding to $e_r = 1.79$ at a bias current of $I_b = 54\,\mu A$ and an applied magnetic flux of $\Phi_a/\Phi_0 = 0.274$. With a flux noise of $S_\Phi^{1/2} = 361\,n\Phi_0/\sqrt{Hz}$ this is a reasonably good result for a conventional, non-optimized dc SQUID.

Its asymmetric counterpart however, exhibits a significantly better performance. The flux noise $S_\Phi^{1/2} = 133\,n\Phi_0\sqrt{Hz}$ is a factor of 2.7 lower. The optimal energy resolution was found at a bias current of $I_b = 56\,\mu A$ and a flux bias of $\Phi_a/\Phi_0 = 0.855$. The extracted values of $\varepsilon = 32\hbar$ and $e_r = 0.52$ were even higher than predicted by simulations. This effect may be accounted for by chaotic switching between voltage states which appears in simulation, whereas the actual device does not seem to be prone to this chaotic behavior [RNM+12].

Conclusion

The simple but novel design, where instead of both junctions, only one of the two is shunted with a resistor $R^{-1} = R_1^{-1} + R_2^{-1}$ proves to be a powerful way to enhance the performance of dc SQUIDs. Simulations based on asymmetric Langevin equations reliably predict the voltage response and noise performance of the devices. The new biasing scheme developed at the PIT II, *University of Tübingen*, allows high precision measurements of the noise performance using the flux-locked loop mode of a SQUID amplifier, without the drawback of a low input impedance. By means of this it has been possible to show, that strong R-asymmetries enhance the normalized energy resolution e_r by a factor of 3.4 as compared to a comparable SQUID with $\alpha_R = 0$. Although the enhanced performance comes at the cost of a very narrow flux bias range, *c.f.* Fig. 3.23 (h), the energy resolution seems to be a rather robust parameter with respect to the bias current I_b, which when altered $\pm 4\,\mu A$ still yielded $e_r < 0.55$ [RNM+12]. Although the damping parameter β_c can still be optimized to further increase the overall performance, the normalized energy resolution of $e_r = 0.52$ is the best value ever measured for a dc SQUID to this day [140].

3.4.2. Fractional Josephson Vortex Devices

In the following, results on devices for the creation of fractional flux quanta (φ-vortices) will be discussed. The basic introduction to the phenomena of fractional Josephson vortices was given in subsection 2.4.2. Most of the measurements shown hereafter have been performed at the PIT II, *University of Tübingen*, using different setups, which will be described later. Characterization of the fabricated chips has been done at the IMS, KIT, in order to identify working samples. Due to the very complex structure of most devices involving fractional vortices, the yield for high-quality devices is lower than for simpler structures such as the above described dc SQUID. The section is divided into several parts, starting with the general characterization based on IVC, $I_c(\Phi)$ and $I_c(I_{inj})$ measurements of linear and annular samples. Next, the resonant excitation of single vortices and vortex molecules in annular LJJs will be discussed, before the section ends with a discussion of the results obtained on linear single- and multi-vortex devices. Investigations of linear vortex molecules initiated the thorough refinement of the Nb/Al-AlO$_x$/Nb technology introduced in section 3.2 to achieve sub-µm structures in all layers. Using this refined process, structures with high critical-current densities and nano injectors were fabricated and tested at mK-temperatures in the quantum regime.

Characterization of Fractional Vortex Devices

Fractional vortex devices investigated within this thesis are based on long Josephson junctions of either linear or annular shape. The characterization after fabrication starts as usual with the measurement of IV characteristics. As mentioned in subsection 3.3.3, the critical current of long junctions may be suppressed strongly if the length exceeds the Josephson penetration depth significantly. To counter this phenomenon, the second generation of linear devices included the newly developed biasing scheme, incorporating a serial resistor and narrow superconducting bias lines. In the case of narrow annular long Josephson junctions, the current density distribution $\iota(z)$ integrated over the coordinate x, matches the bias current profile of a wide superconducting strip, approximated in equation 3.7, almost perfectly [144]. This coincidence makes the new, quite complex biasing scheme unnecessary for annular devices. In Fig. 3.24 (a) the IVC of a linear LJJ is shown with homogenized bias current feed and in (b) the IVC of an annular LJJ [123] is shown, biased with a conventional superconducting strip at $T = 4.2\,\text{K}$. From the IVCs, quality parameters such as the I_cR_N-product, the gap voltage V_g and the R_{sg}/R_N-ratio were extracted. The shown junctions are

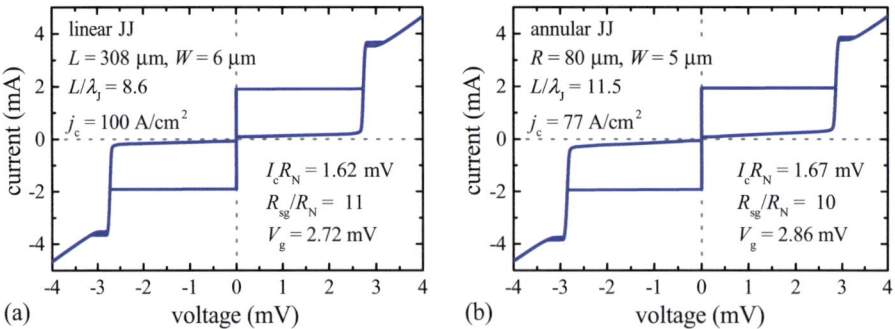

Figure 3.24.: IVC of a linear (a) and annular (b) long JJ. The linear JJ is bias using the novel biasing scheme for homogeneous bias current feed, introduced in subsection 3.3.3, while the annular JJ is biased using a conventional superconducting strip. The quality of both LJJs is sufficient for the investigation of vortex dynamics. (b) Is taken from [123].

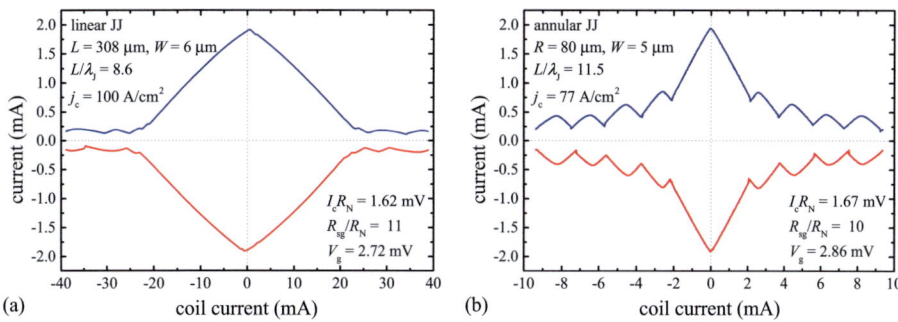

Figure 3.25.: $I_c(H)$ of a linear (a) and annular (b) long JJ. The linear JJ shows slight eigenfield effects as well as trapped flux. The annular devices is free of both anomalies. (b) Is taken from [123].

comparable in j_c, of typical dimensions for $\wp$-vortex devices and exhibit quality parameters suitable for the investigation of vortex dynamics.

Next, the $I_c(H)$ dependences have been recorded for all junctions. By this, any eigenfield effects due to misaligned bias lines or trapped flux due to remanent magnetic stray fields, could be detected. Since long junction devices are extremely susceptible to magnetic fields the setup was equipped with two rf shields and two cryoperm shields. Furthermore, special care was taken during cool-down of the sample, in order to avoid trapping flux upon transition into the superconducting state. This included a very slow cool-down procedure as well as disconnecting the sample from any read-out electronics, while the samples have been shorted during the cool-down procedure. To start measurements, the sample was first connected to the electronics rack before it was un-shorted. Although this procedure minimized the risk of trapping flux, devices exceeding $L/\lambda_J \approx 12$, had to be thermally cycled several times in order to achieve a flux free state.

In Fig. 3.25 (a) and (b) the $I_c(H)$ dependences of a linear and an annular LJJ is depicted, respectively. For the linear device a point symmetric shift of the trace for positive and negative applied bias current is visible, indicating a small eigenfield effect, which is most likely due to a slight misalignment of the wiring electrode to the bottom electrode, effectively causing a minimal fraction of the magnetic field induced by the bias current to penetrate the barrier in $\pm x$-direction. Additionally, the curve is not perfectly symmetric with respect to the vertical bias current axis, indicating that there is trapped flux affecting the structure. However, both of these anomalies are of very small magnitude and may thus be neglected for further investigations. In (b), the $I_c(H)$ dependence of the annular device is depicted. The symmetry with respect to both axes indicated a flux free state. It should be noted that the coil currents are not comparable for the two measurements shown. Therefore a direct comparison between the two $I_c(H)$ dependences with respect to the applied magnetic field / magnitude of the coil current cannot be drawn.

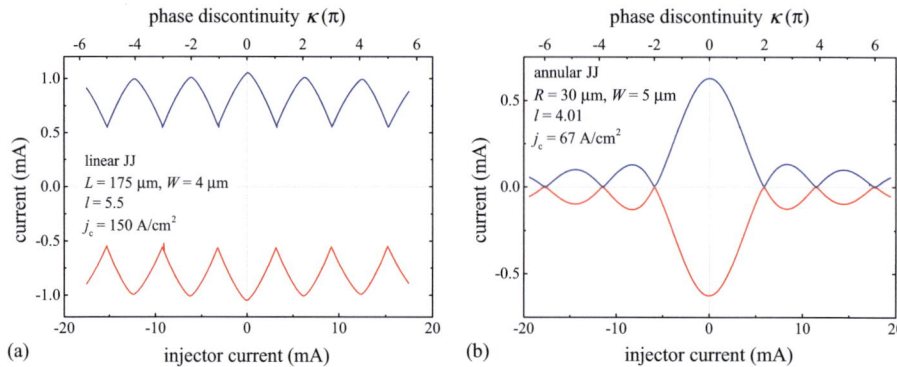

Figure 3.26.: $I_c(I_{inj})$ of a linear (a) and annular (b) long JJ. The critical current of the linear JJ decreases up to an applied injector current corresponding to $\kappa = \pi$, before the vortex flips into a complimentary vortex under the emission of a full fluxon, escaping at the end of the LJJ. For larger κ $I_c(I_{inj})$ rises again, reaching its maxima at $\kappa = 2\pi n$. The $I_c(I_{inj})$ curve of the annular device follows a Fraunhofer pattern, since the vortices cannot escape the system, due to the annular shape of the junction. Data taken from [123].

The last characterization step is typically the $I_c(I_{inj})$ dependence. For interpretation of the characteristics it is important to understand the difference in the boundary conditions for linear and annular devices and the number as well as the position of the injector pairs. Let us first consider devices with a single injector pair. Fig. 3.26 (a) and (b) show the $I_c(I_{inj})$ of a linear and annular LJJ with a one artificial phase discontinuity in the center. Both junctions exhibit extraordinarily high quality parameters and follow ideal $I_c(I_{inj})$ characteristics.

In the case of a linear LJJ, the modulation of the critical current follows a 2π-periodic curve with a slight decrease in magnitude of the side maxima, due to the non-ideal shape of the phase discontinuity. The curve exhibits cusp-like minima at $\kappa_{min} = (2n+1)\pi$ and maxima at $\kappa_{max} = 2\pi n$. The modulation depth is mainly dependent on the normalized length of the JJ, whereas the decrease of the side maxima is mainly determined by the normalized width of the injector pair $W_{pair} = (2W_{inj} + dX) \cdot \lambda_J^{-1}$. A precise derivation of these dependences is given in [83, 145]. The fact that the critical current reaches almost its original value $I_c(\kappa = 0)$ for the $\kappa = 2\pi n$ is due to the open topography of the system. When increasing the injector current past $I_{inj}(\kappa = \pi)$, the direct vortex with $\wp = -\kappa$ changes to an energetically favorable complementary vortex with $\wp = -\kappa + 2\pi \operatorname{sgn}(\kappa)$ under the emission of a full fluxon. The Josephson vortex with $\wp = 2\pi$ exhibits solitonic behavior and may move freely along z in the JJ. Under the influence of the Lorentz force, due to the bias current, the fluxon is accelerated away from the injector-pair towards the end of the LJJ at $\tilde{z} = 0$ or $\tilde{z} = 1$, where it may escape the system. At the artificial phase discontinuity, the complementary vortex remains with a decreasing topological charge $\wp$ up to $I_{inj}(\kappa = 2\pi)$.

For annular junctions the boundary conditions prevent the emission of fluxons due to their

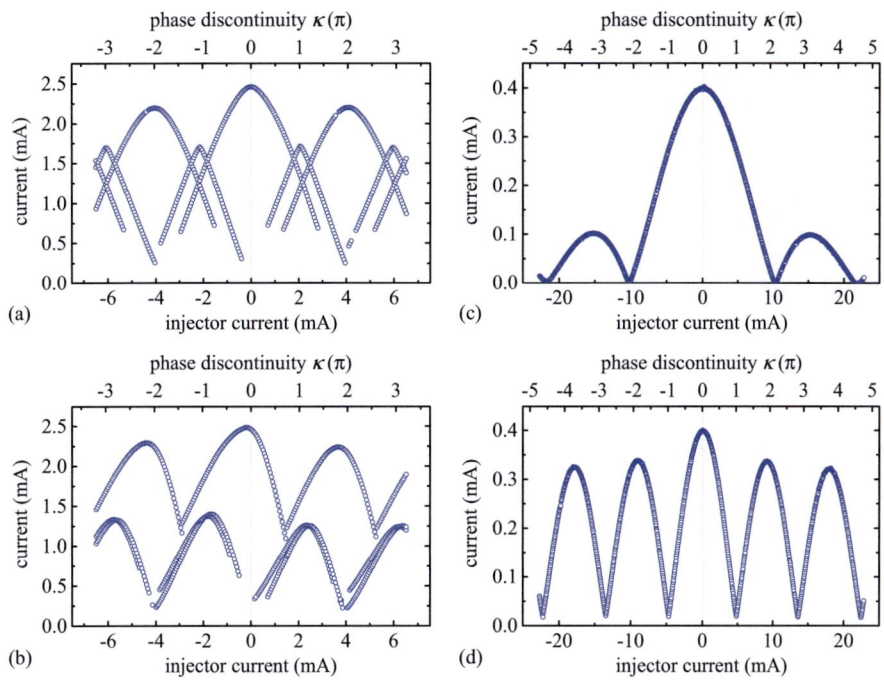

Figure 3.27.: Ferromagnetic (FM) and anti-ferromagnetic (AFM) $I_c(I_{inj})$ dependences of a linear and annular long JJ. (a) Shows the FM case in a linear JJ. The curve is symmetric with respect to the vertical current axis, whereas the AFM case in (b) shows the asymmetry expected from theory. (c) Shows the FM case for an annular vortex molecule. Again, the FM orientation is symmetric with respect to the vertical current axis. The AFM mode in an annular LJJ is shown in (d).

closed topography. Upon emission of a full fluxon at $\kappa = (2n+1)\pi$, the Josephson vortex simply circles the junction once, recombining with the remaining complementary vortex and forms the original direct vortex. Consequently the $I_c(I_{inj})$ characteristic of an annular LJJ follows a Fraunhofer pattern, with steadily decreasing side maxima for increasing $|\kappa|$.

Devices with 2 injector pairs have more degrees of freedom concerning the orientation of fractional vortices. In the following the ferromagnetic (FM) state denotes the case where both artificial phase discontinuities are of equal sign and the anti-ferromagnetic (AFM) state denotes the case of opposite signs. Fig. 3.27 (a) and (b) show the FM and AFM case in a linear JJ with 2 injector pairs symmetrically arranged around its center. The FM characteristics show a symmetric dependence with respect to the vertical current axis, which is in principle comparable to the $I_c(I_{inj})$ curve of a single φ-vortex. The extra traces visible in the vicinity of $\kappa = (2n+1)\pi$ resulting in small maxima, in fact correspond to the state where one of the two vortices has already flipped in its complementary state. Effectively, there the vortices are arranged anti-ferromagnetically, representing the energetically favorable state in this range of $\kappa_{1,2}$.

The AFM excitation of the fractional vortices results in an asymmetric $I_c(I_{inj})$ characteristic. This effect can be explained by the bias current exerting a Lorentz force on the vortices. Due to the opposite sign of the vortices they are either pushed towards or pulled apart from each other due to the Lorentz force. Thus, the asymmetry with respect to the bias current axis is to be expected. For negative bias currents this effect is of course reversed, resulting in characteristics, which are symmetric with respect to the origin.

In the case of an annular LJJ with a normalized length of $l \approx 2.7$, both FM Fig. 3.27 (c) and AFM (d) characteristics are symmetric. Due to the annular topography the vortices are in a quasi-infinite system, *i.e.* each vortex neighbors the other vortex on both sides. Effectively the FM mode (c) is identical to that of a single vortex $I_c(I_{inj})$ dependence, with the exception that the minima are not located exactly at $2\pi n$, but at a value $\kappa_{min,n} \gtrsim 2\pi n$ (for the nth minimum). This is an effect of the finite dimensions of the real device [86]. In the case of an anti-ferromagnetic orientation of the vortices (d), the characteristics change into those of a single vortex characteristics of a linear LJJ. This effect can be explained by the absorption fluxons emitted by the other vortex. Effectively, this means that emitted fluxons recombine with the other $\wp$-vortex, which results in the same effect as escaping from the system as is possible in for linear LJJs. Again, the minima are not exactly at $\kappa = (2n+1)\pi$ due to the finite dimensions of the real device [83]. The conversion factor of the injector current to the phase discontinuity has been extracted from previous measurements on each single $\wp$-vortices.

3.4.3. Resonant Activation of Fractional Vortices

Once the characterization has been completed and high-quality devices have been identified, more complex and time consuming measurements have been undertaken. One particular focus of this work included the investigation of the resonant excitement of fractional vortices by external microwave radiation. Unlike conventional Josephson vortices, fractional $\wp$-vortices are pinned to their point of origin, *i.e.* the injector pair. Due to this pinning they may oscillate around the resting equilibrium state [85, 88, 145–148][KMB$^+$12]. When applying a constant bias current, the vortex will be deformed under the Lorentz force. When released of this force, *i.e.* the bias current is suddenly reduced to $I_b = 0\,\text{mA}$, the vortex will oscillate with a frequency [149]:

$$\omega_0(\wp) = \omega_{pl} \sqrt{\frac{1}{2}\cos\frac{\wp}{4}\left(\cos\frac{\wp}{4} + \sqrt{4 - 3\cos^2\frac{\wp}{4}}\right)}. \qquad (3.16)$$

Eq. (3.16) has been derived for infinitely long junctions with a damping parameter $\alpha = 0$. The oscillatory eigenfrequency of a single fractional vortex in a strongly underdamped junction approaches the plasma frequency introduced in Eq. (2.29) for $\wp \to 0$ and vanishes for $\wp \to 2\pi$.

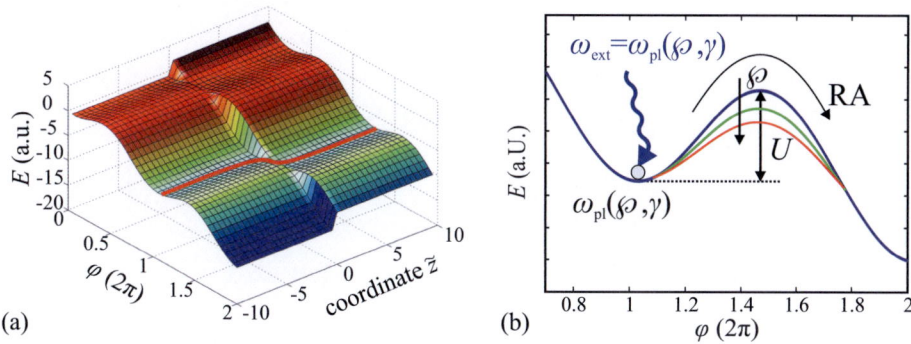

Figure 3.28.: (a) Shows the extended tilted washboard potential for a LJJ with one artificial phase discontinuity in the center. In (b) this scenario is mapped to a single phase particle with an effective potential barrier to the right, which is dependent on the topological charge of the phase discontinuity $\wp$.

For $\gamma \neq 0$,

$$\omega_0(\wp, \gamma) \approx \omega_0(\wp, 0) \left[1 - \left(\frac{\gamma}{\gamma_c} \right)^2 \right]^{1/4} \tag{3.17}$$

is a good approximation [81, 86, 87, 147]. Here the vortex depinning current is given by $\gamma_c(\wp) = \left| \frac{\sin(\wp/2)}{\wp/2} \right|$, for $|\wp| \leq 2\pi$. An analytical solution for $\omega_0(\wp, \gamma)$ is not known, however Eq. (3.17) deviates from the numerical solutions merely by a few percent for $\gamma \to \gamma_c$ and is exact for $\wp \to 0$.

Phenomenologically, this results in a decreased critical current of the fractional vortex device for $\wp \neq 0$. Additionally, the eigenfrequency is reduced as compared to the plasma frequency of a plain JJ. In terms of the tilted washboard potential introduced in section 2.2, the phase in a long JJ with an artificial phase discontinuity can be described by a potential extended along the coordinate z with an abrupt shift in the phase on a length scale of W_{pair} at the point of the discontinuity. The phase is then described by a chain of interlinked point-like phase particles which may vary on a characteristic length scale λ_J. In Fig.3.28 (a) the extended tilted washboard potential for a $\wp$-vortex device with a single artificial phase discontinuity in the center is shown. The phase chain is depicted as a red line. In the vicinity of the injector pair the phase sees an effective decrease of the potential barrier, which results in a locally reduced switching current. If the chain overcomes this reduced barrier locally, it may drag the remaining chain along, switching the entire junction. In Fig. 3.28 (b) this scenario is mapped to a single phase particle in a classical washboard potential with a barrier U to the right, varying with changes in $\wp$. By exciting the $\wp$-vortex with an external microwave radiation, the particle may escape from the potential well, even for $\gamma < \gamma_c$.

Experimentally, the effective barrier height U can be extracted by resonant microwave

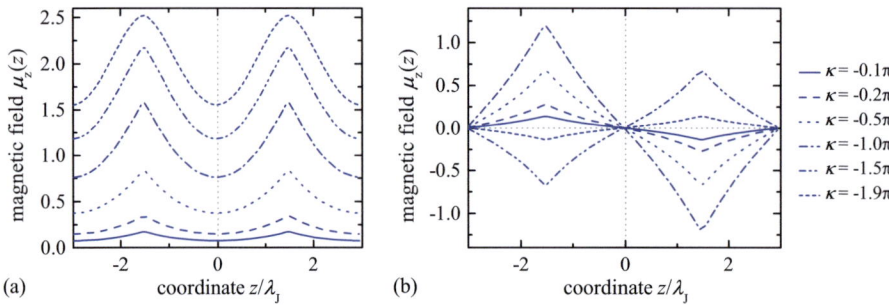

Figure 3.29.: (a) Shows the continuous magnetic field for the sFM orientation in an annular vortex molecule of length $l = 6$ for different values of κ. (b) Shows the corresponding sAFM orientation of the molecule. Numerical simulation have been done by Dr. E. Goldobin at the *University of Tübingen*, using the software StkJJ [150].

excitation of the fractional vortex. The measurement technique is comparable to the one employed for extraction of the Josephson plasma frequency [85]. Switching histograms with typically 10000 points have been measured on the device under the influence of ω_{ext}. For a given applied frequency $\omega_{ext} = \omega_0(\wp, \gamma)$, two peaks in the histogram will appear. The peak at higher bias currents corresponds to the resonant activation of the JJ itself, *i.e.* γ_c, whereas the lower peak at γ_{res} corresponds to the resonant excitement of the $\wp$-vortex. Since the JJ itself acts as a nonlinear oscillator, the resonances will shift towards lower frequencies when high microwave (MW) power is applied. Particularly for low values of γ, where the potential barrier U is still large and thus high MW power is required, this effect leads to deviations in the extractions of $\omega_{pl}(\gamma \to 0)$. Therefore, γ_{res} has been determined at the lowest possible applied MW power [KMB$^+$12]. The measurements have been performed at LHe-temperature, while the sample has been mounted inside a copper box for rf shielding. For magnetic shielding, a single cryoperm shield has been used. The microwave signal has been applied via a semi rigid cable and capacitively coupled to the device.

Even though numerous such measurements have been performed within this thesis [85, 88], here the focus will be on the investigation of vortex molecules [KMB$^+$12]. In the following, the particular case of annular junctions with two fractional vortices at a distance a from each other will be discussed. Here $a = 1/2$, where $l = 2\pi R$ and R is the mean radius of the annular LJJ. In such a structure, with both vortices having the same absolute value of $|\kappa|$, there are four irreducible vortex arrangements [151]. For signs of the phase discontinuities being equal $(-\kappa, -\kappa)$ there is the symmetric parallel (sFM) molecule with $(\wp_1, \wp_2) = (+\kappa, +\kappa)$ and the asymmetric anti-parallel (aAFM) molecule with $(\wp_1, \wp_2) = (+\kappa, +\kappa - 2\pi)$. In the case of opposite signs $(+\kappa, -\kappa)$, there is the symmetric anti-parallel (sAFM) molecule with $(\wp_1, \wp_2) = (-\kappa, +\kappa)$ and the asymmetric parallel (aFM) molecule with $(\wp_1, \wp_2) = (-\kappa + 2\pi, +\kappa)$. Although all of these arrangements are stable, the detailed investigation was limited to the symmetric cases (sFM and sAFM)

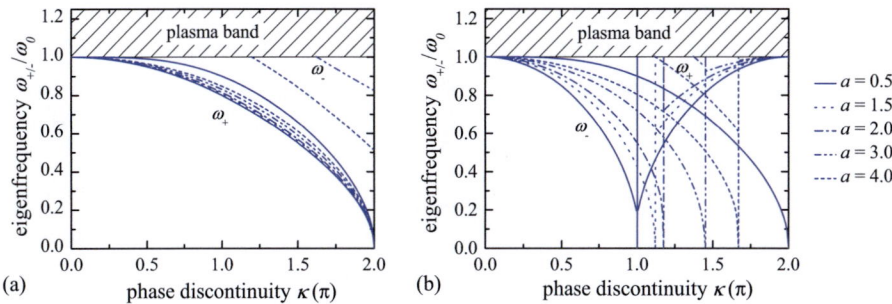

Figure 3.30.: (a) Shows the eigenmodes for the sFM orientation of molecules of different lengths. In (b) the corresponding plot is shown for the sAFM orientation of such molecules.

for reasons of simplicity. For both cases the continuous magnetic field $\mu_z(z)$ [16] is plotted without any bias current applied in Fig. 3.29 (a) and (b) respectively. In the case of the sAFM orientation there is a critical value of the phase discontinuity κ_c ($\pi < \kappa_c < 2\pi$)) which mainly depends on the distance a between the two injector pairs. For values of $\kappa < \kappa_c$ the direct vortices $(\wp_1, \wp_2) = (-\kappa, +\kappa)$ are stable. Above κ_c this configuration becomes instable due to the increasing interactive force between the two vortices. Thus, for $\kappa > \kappa_c$ the two fractional vortices exchange a full fluxon and the sAFM case flips into the complementary counterpart with $(\wp_1, \wp_2) = (+\kappa - 2\pi, -\kappa + 2\pi)$. Consequently there two stable states for values $(2\pi - \kappa_c) < |\kappa| < \kappa_c$, which however have different energies [151][KMB+12].

Due to an upper frequency limitation of the measurement setup, the investigations were limited to the two lowest eigenmodes of the vortex molecule. The in-phase oscillation of the vortices is denoted with ω_+, while the out-of-phase oscillation is denoted with ω_-. Fig. 3.30 (a) shows the dependences of the normalized eigenfrequencies for molecules of different lengths l in case of the symmetric FM orientation. In this particular case the in-phase oscillation ω_+ of the vortex molecule exhibits a lower frequency than the out-of-phase oscillation ω_-. In (b) the eigenfrequencies $\omega_\pm$ are depicted for the sAFM case. Unlike in the previous example, here $\omega_- < \omega_+$. The vertical lines mark the critical values of $\kappa = \kappa_c$ at which the vortices change from their direct states to their complementary states. As can be seen, the behavior of the eigenfrequency changes due to a change in the topological charge.

As could be expected from the behavior shown in Fig. 3.30, the eigenfrequencies of the vortex molecule split significantly for the different modes (FM and AFM) with varying distance a between the two artificial phase discontinuities. Fig. 3.31 (a) shows the splitting for $(|\kappa_1| = 0.6\pi, |\kappa_2| = 0.6\pi)$ for different distances a. As can be seen, the eigenfrequencies ω_- and ω_+ deviate strongly with decreasing junction length, whereas the splitting is marginal for very long junctions. The solid circles depict measured values of the lowest (first) eigenfrequency for three different samples. The second eigenmode could only be detected for the largest samples (open circles). This is most probably due to the fact that this

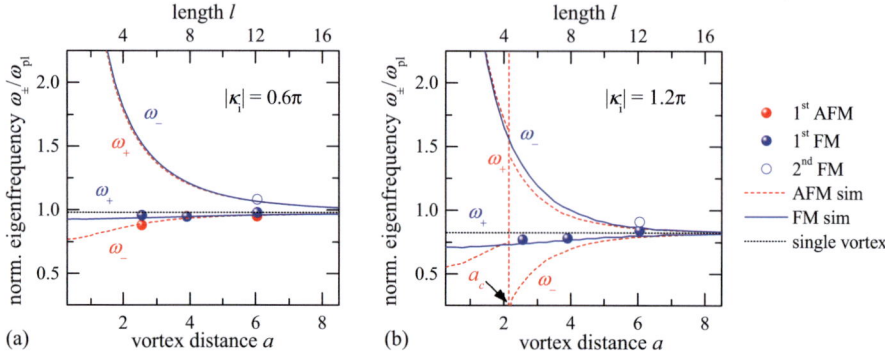

Figure 3.31.: Splitting of eigenmodes for annular molecules with varying a. (a) Shows the simulations and measurement results for $\kappa = 0.6\pi$ and (b) for $\kappa = 1.2\pi$. Solid (dashed) lines correspond to the simulation of the FM (AFM) configuration. For comparison, the eigenfrequency of a single vortex in an infinitely long JJ is also shown (dotted lines). Full symbols correspond to fit values of the first eigenmode from experimental data of samples and open symbols to the second eigenmode.

particular sample exhibited the lowest critical-current density $j_c \propto \omega_{pl}^2$. In (b) the case for $(|\kappa_1| = 1.2\pi, |\kappa_2| = 1.2\pi)$ is shown. Again the splitting is clearly visible and th measurements coincide well with the simulation.

In Fig 3.32 the resonance scans for the FM (a) and AFM (b) mode are shown for the longest investigated sample ($l \approx 12.1$). The resonance peaks coincide with the numerical simulations (dashed lines) in all cases, except for ω_+ in the AFM mode. This effect has been observed for all investigated annular vortex molecule devices, the origin however is unclear. It should be noted, that for reasons of clarity, parasitic resonances due to thermal excitations at microwave powers as well as resonances attributed to the measurement setup are not shown in Fig. 3.32. In (c) and (d) a comparison is drawn to samples of smaller lengths $l \approx 5.1$ and $l \approx 7.8$. As expected the length dependence is small in case of a FM molecule orientation (c), whereas the effect is more dominant for the AFM mode (d).

Conclusion

From the experiments involving artificial fractional-vortex molecules in annular Josephson junctions it can be concluded, that the Nb/Al-AlO$_x$/Nb fabrication process is suitable for highly complex structures involving long Josephson junctions and μm sized current feeds for creation of fractional flux quanta. Spectroscopic measurements have shown that the splitting of the oscillatory eigenmodes depends on the vortex distance a, the topological charge $\wp_i$ of the individual vortices, as well as the applied bias current γ. Furthermore, the molecule constellation, *i.e.* FM or AFM orientation, strongly affects the resulting eigen-

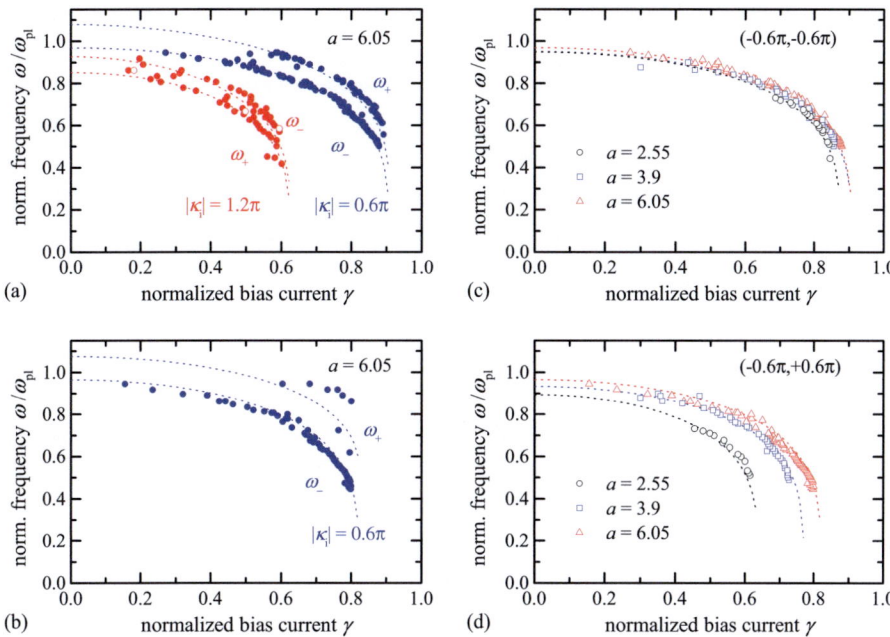

Figure 3.32.: Resonance scans of for different values of κ in the FM (a) and AFM (b) mode of the sample with $l \approx 12.1$. (c) And (d) show the comparison of the resonance scans to shorter samples ($l \approx 5.1$ and $l \approx 7.8$).

frequencies. Numerical calculations reliably predict the eigenfrequencies. In summary, the conducted experiments allow identification of the vortex states in a symmetric molecule.

3.4.4. Linear Vortex Molecules

In the previous section the read-out of vortex states by means of spectroscopic measurements was introduced. While this procedure is certainly a reliable possibility for determining the state of the molecule, it is rather time consuming and therefore not suitable for fast read-out procedures. An alternative approach involves dc SQUIDs placed close to the phase discontinuities, *i.e.* the injector pairs [152]. The high sensitivity of properly designed SQUIDs allows the detection of the polarity of the vortex state, by measuring the magnetic field of the direct or complementary $\wp$-vortex.

The goal of this task of the thesis was the development of linear LJJ 2-vortex molecules with read-out SQUIDs coupled to the $\wp$-vortices with flux transformers. In order to keep the dissipated energy on chip as small as possible the SQUIDs had to be designed as unshunted devices with a current-asymmetry α_I to avoid an extra flux bias of the interferometers. Such linear LJJ vortex molecule devices were predicted to be promising candidates for classical information storage devices (bits) as well as qubits [153, 154]. In the following the developed devices will be discussed, along with measurements of the linear vortex molecule at $T = 4.2\,$K and $T = 300\,$mK.

The geometry of the fabricated samples corresponds closely to the normalized lengths of the anticipated qubit devices as introduced in [154, 155] including the injector pair spacing of $a \approx 1.7$. Such qubit devices are based on the idea of creating a double-well potential where the molecule state may tunnel from the $\uparrow\downarrow$ state to the $\downarrow\uparrow$ state [153, 154]. The fabrication of such devices based on SIFS or Nb-YBCO technology is extremely complicated, due to the fixed topological charge of semifluxons, resulting in very tight requirements for the distance a between the vortices. The fabrication based on Nb/Al-AlO$_x$/Nb technology implementing φ-vortices results in an extra degree of freedom, since the topological charge of the vortices may be tuned at will between 0 and 2π, thus altering their interaction. This relaxes the tight restrictions on the distance a between the injector pairs, since the coupling between the vortices and thus the energy barrier between $\uparrow\downarrow$ and $\downarrow\uparrow$ may be tuned not only by a but also by varying the injector currents. In [155] it has been shown that the acceptable range for the distance between the two injector pairs ranges from $a = 1.57$ to $a = 1.76$. This corresponds to a relaxation of the precision restriction of one order of magnitude as compared to a qubit device based on SIFS or Nb-YBCO technology where the length of the π-regions needs to be $\pi/2 \leq a \lesssim \pi/2 + 0.02$ to observe quantum tunneling [154]. In the following, the design and results obtained on the first generation of linear vortex molecules for qubit application will be discussed. Requirements for the later generations will be outlined and the results on the new single-vortex devices measured at mK-temperatures finalizes this section.

An example of the investigated structures is shown in Fig. 3.33 (a-c), depicting images of the sample during fabrication, *i.e.* before deposition of a second SiO insulation layer and definition of the aluminum flux transformers in (a) and (b) and the final topography in (c). In (a) an overview over the structure is shown with all electrical connections indicated in color. The bias circuitry is shaded in blue with the serial resistor highlighted in light brown. In order to be able to extract an *IVC* without the effects of the serial resistor, a separate voltage pad is connected directly to the JJ. The junction itself is highlighted in bright yellow at the top of figure (a). The two injector pairs are indicated in green, situated atop of the bottom electrode. The distance dZ between the injectors is identical to the junction width $W = dZ = 6\,\mu$m to ensure a one dimensional current flow along z. The connections for the SQUID read-out are highlighted in light red.

Fig. 3.33 (b) shows an enlarged view of the central JJ area together with injectors and the SQUIDs. The distance between the two injector pairs is $a \approx 58\,\mu$m. As can be seen the SQUIDs are designed to have the least eigenfield effects possible, with the bias lines lying right on top of each other. Additional gradiometric designs were fabricated [Bue11], but are not shown here.

The shown samples were fabricated in the conventional fabrication process, *c.f.* sec-

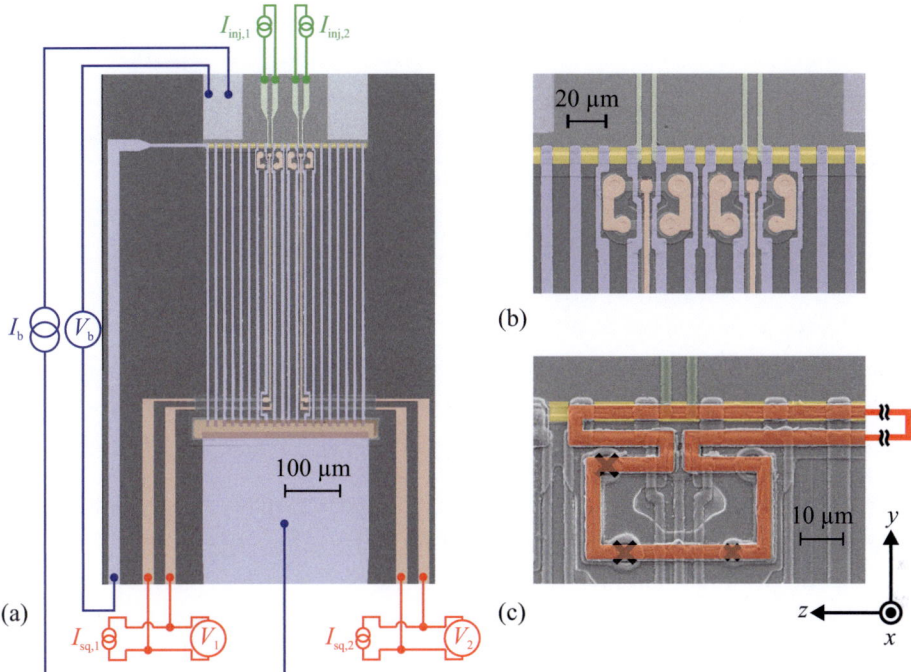

Figure 3.33.: SEM images of a linear vortex molecule with dc SQUID read-out and striped bias lines. (a) Shows the entire structure with all connections indicated for complete measurements. (b) Shows the zoom to the dc SQUID read-out and the injectors. The bias lines are highlighted in blue, the injectors in green, SQUIDs in red, the Josephson junction in yellow and the serial resistor in brown. (c) Shows close-up of a single SQUID and the corresponding injector with the aluminum pick-up loop highlighted in dark red.

tion 3.1, where an additional second insulation layer and a passive Al-layer for pick-up loops (PULs) were integrated using specific thick resists [Bue11]. Due to the large thicknesses of the wiring layer in the conventional process of $d_{\text{w}} \approx 400\,$nm the second insulation layer had to be thicker than the maximum resulting step height, occurring at the intersection of the wiring layer crossing the via to the JJ top electrode. At this point the combined thickness of the first insulation layer and the wiring layer results in step max $\approx 700\,$nm, resulting in a minimum thickness of $d_{\text{I2}} \geq 700\,$nm. Consequently, also the dc sputtered Al for the PULs had to be of a minimum thickness of $d_{\text{PUL}} \geq 900\,$nm to ensure a closed superconducting loop. For reliable patterning, resists with a thickness of $d_{\text{resist}} \approx 3\,\mu$m were used in an inverse lithography and subsequent lift-off. After thorough optimization of the exposure time as well as the pre- and post-baking time, PULs with $3\,\mu$m minimum lateral dimension could be patterned. The final structure is shown in Fig. 3.33 (c) with the additional aluminum pick-up loop highlighted in dark red. The loop closes on the right at an approximate distance of $2\lambda_{\text{J}}$ away from the center of the corresponding fractional vortex, *i.e.* the injector pair. The JJs in the SQUID are indicated by black crosses. The PUL is directly above the

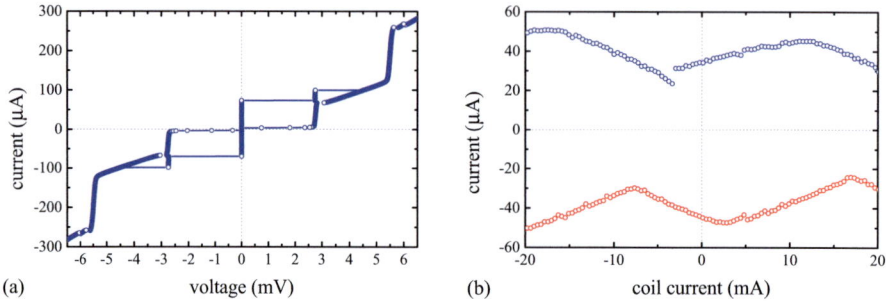

Figure 3.34.: Characteristic measurements of a current asymmetric 3-junction SQUID. (a) Shows a typical IVC and (b) an $I_c(\Phi_a)$ measurement. The curves have been measured on two different devices.

JJ with one arm, such that the magnetic flux of the $\wp$-vortex couples to the superconducting loop and thus increases the coupling to the SQUID.

For testing of the functionality, characteristic measurements as described in the beginning of this section were performed. For the linear vortex molecules, typical critical-current densities were $j_c \approx 100\,\text{A/cm}^2$. The devices with SQUID read-out have minimal lateral dimensions of $2\,\mu\text{m}$ for the injector width and $4\,\mu\text{m}$ for the junction width. The JJ lengths varied between $277\,\mu\text{m}$ and $308\,\mu\text{m}$.

In addition to the characterization of the LJJ, the SQUIDs for the vortex read-out had to be characterized at $T = 4.2\,\text{K}$ preliminary to the full vortex measurements. Fig. 3.34 (a) shows a typical IVC of a current-asymmetric SQUID. The two serially connected junctions in the first SQUID arm are twice the size of the single junction located in series to a via in the second arm. This results in an equal resistance in both arms, however with a current-asymmetry of $\alpha_I = 1/3$, *i.e.* $I_{c,1} = I_{c,2} = 2I_{c,3}$. The unequal current in the two arms results in an effective magnetic field coupled to the SQUID loop even for $\Phi_a/\Phi_0 = 0$, rendering an extra flux bias obsolete. Fig. 3.34 (b) shows a $I_c(\Phi_a)$ measurement of a 3-junction SQUID using a RTA. As can be seen, the maximum critical current is reached only for an applied coil current $I_{coil} \neq 0\,\text{mA}$, *i.e.* $\Phi_a/\Phi_0 \neq 0$. In the experiment, the flux Φ_a to be detected, is created by the artificial vortex induced by $I_{inj,k}$, where $k = 1,2$ denotes the two injector pairs. In order to keep the necessary current for the creation of a $\wp$-vortex at a minimum, the injectors had no connection in the wiring layer, forcing the injector current to flow through the top trilayer electrode and therefore very close to the barrier, *c.f.* Fig. 3.33 (b). The vortex creation is further supported by the increased inductance of the top trilayer electrode as compared to the rather thick wiring electrode.

The critical temperature of the aluminum, deposited for the pick-up loops, is approximately $T_{c,Al} \approx 1.14\,\text{mK}$. Thus there is no increase in coupling efficiency of the vortex to the SQUID during measurements at liquid helium temperature. Fig. 3.35 shows the SQUID

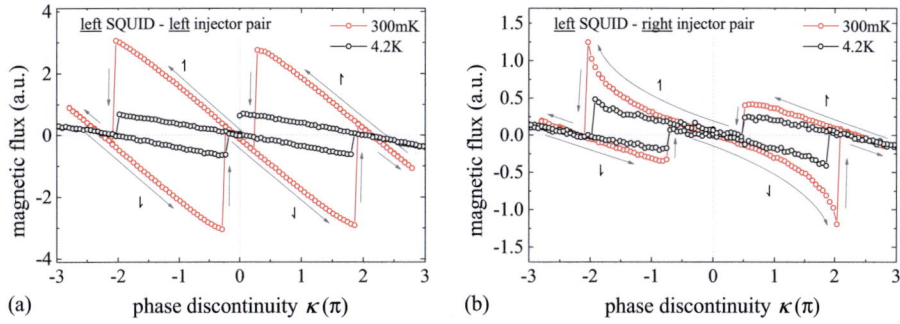

Figure 3.35.: Single vortex SQUID read-out of linear vortex molecule at $T = 4.2\,\text{K}$ (black traces) and $T = 300\,\text{mK}$ (red traces). (a) Shows the response of the left SQUID when a $\wp$-vortex is induced in the left injector pair, while (b) shows the response in the same SQUID when a vortex is induced using the right injector pair. The gray arrows indicate the path of the hystereses. The vortex states are indicated by the black harpoons.

read-out of for the two different vortex excitations. The black traces correspond to the measurements performed at $T = 4.2\,\text{K}$. For clarification the $\partial I_c(\Phi)/\partial\Phi$-slope of the SQUID has been subtracted from the original data, resulting in a current-response in arbitrary units plotted over the phase discontinuity κ. The conversion of the applied injector currents to κ has been extracted by $I_c(I_{inj})$ measurements as described earlier. Each point corresponds to the mean value of the SQUID's switching current extracted from a histogram over 1000 individual I_c measurements. The measurements were performed using an automated software and a current-ramp generator [85, 145][Bue11].

Fig. 3.35 (a) shows the results of the left SQUID response to a $\wp$-vortex induced in the left injector pair. Starting from the origin the injector current was ramped upwards and for each point the mean switching current of the SQUID was extracted. Initially a direct $\wp$-vortex ($\downarrow$) is induced until it flips into the complementary $\wp$-vortex ($\uparrow$). When decreasing $I_{inj,left}$, the vortex remains in its complementary state until it flips back to the direct state near the origin. The resulting hysteretic behavior can also be observed in the reversed case with an injector current of opposite sign. Again, the direct $\wp$-vortex ($\uparrow$) is initialized, flipping to its complementary state ($\downarrow$) close to $\kappa = 2\pi$, describing a similar hysteresis as for positive injector currents. It should be noted that there is no bias current applied in to the long junction, which is why the vortex does not experience any Lorentz-force and thus does not flip to its complementary counterpart at $\kappa = \pi$, but remains to be a direct vortex until it almost reaches the topological charge of a full fluxon $\wp = 2\pi$. Fig. 3.35 (b) shows the response of the identical SQUID to the vortex created in the opposite (right) injector pair. Again, comparable behavior may be observed. Most strikingly, the I_c modulation of the SQUID is only slightly less pronounced (approximately a factor of 0.7) than in the first case (a). Although this difference is definitely possible to detect, it requires a certain minimum averaging to distinguish between the two vortices.

In order to get a stronger signal of the $\wp$-vortex, additional measurements were performed at $T = 300\,\mathrm{mK} < T_{\mathrm{c,Al}}$. In Fig. 3.35 (a) and (b) the results are depicted in red. In the superconducting state of the PULs, they act as flux transformers, collecting the magnetic flux of the fractional vortex at the point of its origin, and coupling it to the SQUID due to flux conservation in superconducting ring structures. In the case of a vortex induced by the left injector pair, the signal magnitude increases by a factor of approximately 4.3 as compared to measurements at $T = 4.2\,\mathrm{K}$. It should be noted that this effect is supported by the increase in the critical current of the SQUID when cooled from $T = 4.2\,\mathrm{K}$ down to $T = 300\,\mathrm{mK}$. However, this increase can be estimated to be less than 10 % and therefore cannot be the sole explanation of the 430 % increase in signal magnitude.

Whereas an almost linear amplification of the SQUID response to the $\wp$-vortex can be observed in figure (a), there is a strong non-linear effect visible in the case of the opposite vortex excitation in figure (b). Although this effect is not yet fully understood, it can most probably be explained by the ratio of the physical vortex size to the distance between the neighboring PULs. When the right vortex grows in size, it may overlap with the region enclosed by the left PUL when exceeding a certain threshold. In this case, magnetic flux will be coupled not only to the right, but also to the left PUL. With increasing κ this coupling will also increase, resulting in a pronounced response in the read-out of the left SQUID. Final proof of this hypothesis requires additional designs of linear vortex devices with varied distances between the neighboring pick-up loops. Despite this parasitic effect, the difference between the coupling strength of the two vortices is sufficient to easily distinguish between them. For small values of κ the left SQUID rarely detects any increased signal of the right vortex as compared to measurements at $T = 4.2\,\mathrm{K}$ and still exhibits a signal 2.5 times larger for the left vortex than for the right vortex at $\kappa \lesssim 2\pi$. Finally, it should be noted that the discrepancy of the switching point in the hystereses, stems from the fact that the values of κ have been normalized to previously measured $I_{\mathrm{c}}(I_{\mathrm{inj}})$ measurements. During measurements of the shown flux responses of the SQUID to induced $\wp$-vortices, small amounts of flux may have been trapped in the LJJ, causing a shift in the I_{inj}-to-κ-conversion. Such effects however do not affect the qualitative results.

Fig. 3.36 shows the read-out of vortex molecules in the (a) $0 - \kappa - 2\kappa$ and (b) $0 - \kappa - 0$ states. Again, the black traces correspond to measurements at $T = 4.2\,\mathrm{K}$, whereas the red traces correspond to $T = 300\,\mathrm{mK}$. Similarly to the case of a single vortex described above, the read-out signal at $T = 300\,\mathrm{mK}$ is again much stronger than at 4.2 K, due to the superconducting flux transformers. As has been derived in [155], both the $0 - \kappa - 2\kappa$ as well as the $0 - \kappa - 0$ constellation may be used to initialize the vortex to the desired state for investigation of vortex tunneling. Experimentally it turns out that ramping in the $0 - \kappa - 2\kappa$ state is preferable, since it may easily be tuned to the desired intermediate states of direct and complementary vortices.

Measurements depicted in Fig. 3.36 (a) show asymmetries in the measurements that

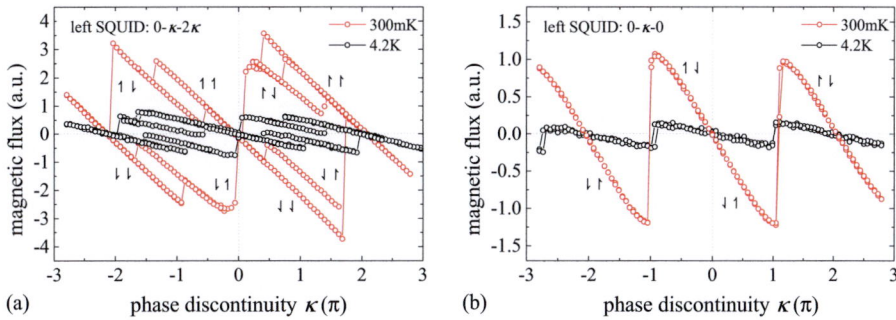

Figure 3.36.: Two vortex SQUID read-out of linear vortex molecule at $T = 4.2\,\mathrm{K}$ (black traces) and $T = 300\,\mathrm{mK}$ (red traces). (a) Shows the response of the left SQUID when the two $\wp$-vortices are induced in the $0 - \kappa - 2\kappa$ constellation, while (b) shows the response in the same SQUID when the $0 - \kappa - 0$ state is induced. The individual molecule states are indicated along the traces.

are not yet fully understood, however are believed to be due to slight inhomogeneities in the LJJ, causing increased pinning of the fractional vortices under some circumstances. Nevertheless, from the measurement at liquid helium temperature in (a) the distance between the injector pairs can be estimated to be $a \approx 1.7$ according to the transition from $\uparrow\uparrow$ to $\downarrow\uparrow$ [151]. This meets the requirements for a fractional vortex based qubits as derived in [155] very well. Although the device under test meets the geometrical and electrical requirements to potentially observe quantum tunneling, the energy barrier remains too large for tunneling effects. As can be seen in (a) the difference in the measured flux $\Delta\Phi_{\mathrm{qb}} = \left|\Phi_{(\downarrow\uparrow)} - \Phi_{(\uparrow\downarrow)}\right| = \left|\Phi_{(\uparrow\downarrow)} - \Phi_{(\downarrow\uparrow)}\right|$ remains almost constant over the entire range of the existence of these intermediate states. Being a measure for the energy barrier $\Delta\mathscr{U}$, this result is contradictory to the theoretical expectations, which predict that $\Delta\mathscr{U} \to 0$ for $\kappa \to 0$ or $\kappa \to 2\pi$, under the assumption that $a > a_{\mathrm{c}} = \pi/2$ [155].

Samples with high critical-current densities were also fabricated using the refined self-planarized process and tested for their crossover temperature to the quantum regime. In [47] it has been shown, that the crossover temperature is also dependent on the junction size. This effect lowers T^* somewhat, which is why trilayers with $j_{\mathrm{c}} \approx 3\,\mathrm{kA/cm^2}$ were chosen for fabrication. For short JJs this should result in a crossover temperature of about $T^* \approx 1.56\,\mathrm{K}$ according to Eq. (2.30). Fig. 3.37 (a) shows the dependence of the standard deviation extracted from switching histograms of 1000 measurements over the bath temperature for a LJJ. The lateral dimensions of the junction are $L = 40\,\mathrm{\mu m}$ and $W = 0.83\,\mathrm{\mu m}$. This results in a critical current of $I_{\mathrm{c}} = 1.1\,\mathrm{mA}$ at $T = 4.2\,\mathrm{K}$. The IVC is depicted in the inset. Blue open symbols indicate the classical regime, where the escape of the phase particle is dominated by thermal excitation. The quantum regime is indicated by green open symbols. Here, the activation mechanism is dominated by macroscopic quantum tunneling (MQT) as described in section 2.2 and [47, 49–51]. The crossover temperature is clearly below expec-

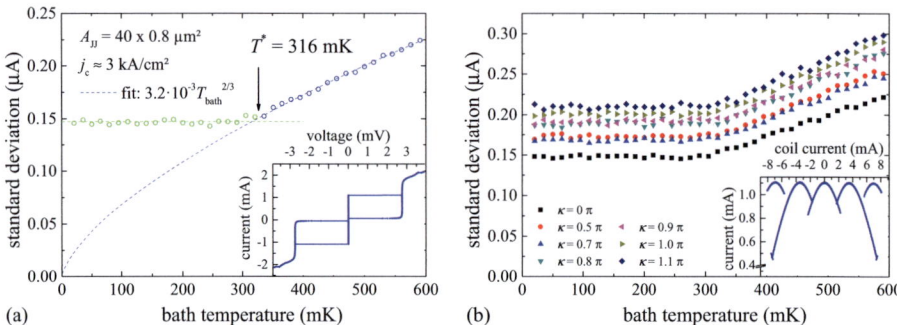

Figure 3.37.: In (a) the temperature scan of σ for a LJJ is shown. The open blue symbols indicate the classical regime, whereas the quantum regime is indicated by open green symbols. The extracted crossover temperature is $T^* = 316\,\text{mK}$. The inset shows the IVC. In (b) the dependence of σ on the size of a fractional vortex is shown. The inset depicts the $I_c(I_{inj})$ characteristic.

tations for such a high j_c, due to the large JJ area and the fact that the Josephson phase has an additional degree of freedom. Nevertheless, it is sufficient for the investigation of the κ dependence of the observed MQT in the available dilution fridge with a base temperature of $T_{base} \approx 25\,\text{mK}$.

The junction is equipped with one injector pair in the center, having lateral dimensions of $W_{inj} = 0.6\,\mu\text{m}$ and $dZ = 0.8\,\mu\text{m}$. Fig. 3.37 (b) shows temperature scans from $T_{bath} = 600\,\text{mK}$ down to $T_{bath} = 25\,\text{mK}$ for various different values of κ. The inset shows the $I_c(I_{inj})$ dependence of the LJJ. For reason that are still unknown, the $I_c(I_{inj})$ characteristic seems to be π-periodic instead of 2π-periodic, resulting in a decreased modulation depth. This is also believed to be reason, why σ is not fully proportional to changes in κ. Nevertheless, the transition to the quantum regime could be observed for all applied κ. Combined with the fact that by tuning of κ the crossover to the quantum regime can be tuned, these results show promise for future device generations.

Conclusion

The above shown measurements have demonstrated that the read-out of single vortex states as well as two vortex molecules using current-asymmetric unshunted SQUIDs is a fast and reliable alternative to the previously discussed spectroscopic measurement. In order to clearly distinguish the two vortices it needs to be ensured that one of the vortices is coupled more strongly to the SQUID than to the other. This was achieved by the use of superconducting flux transformers, which were realized using aluminum pick-up loops in an extra metallization layer [Bue11]. Due to the lower critical temperature of Al than that of Nb, it was possible to show the influence of the PULs by a comparison between measurements at $T = 4.2\,\text{K}$ and $T = 300\,\text{mK}$. Although the investigated structures met the predicted geometrical and electrical requirements for the observation of quantum tunneling of the intermediate $\wp$-molecule states, the energy barrier $\Delta\mathcal{U}$ remained too large for further experiments. Consequently, measurements of these structures at $T \lesssim 25\,\mathrm{mK}$ have not been carried out. Why $\Delta\mathcal{U}$ remains almost constant over varying κ is still part of ongoing investigations. In order to minimize the effort with respect to the measurement equipment for future investigations involving fractional vortices in the quantum regime, new geometrical and electrical requirements were derived. This included a significant increase in the critical-current density, making strongly decreased lateral dimensions necessary to keep the critical current of the LJJ below approximately $1\,\mathrm{mA}$. The current-limitation stems from the limitations in cooling power of the dilution fridge available at the PIT II. This led to the refinement of the fabrication process as described in section 3.2, including sub-μm injectors. First measurements of standard deviation of the switching histograms at varying temperature, showed a significant increase of the crossover temperature up to $T^* \approx 300\,\mathrm{mK}$.

3.4.5. Improved Flux-Flow Oscillators

In collaboration with the *Institute of Radioengineering and Electronics* (IRE) of the *Russian Academy of Sciences* (RAS), Moscow, Russia and the *Physics Institute* at the *Technical University of Denmark* (DTU), Lyngby, Denmark, a modification of the conventional flux-flow oscillator (FFO) has been developed within this thesis. The FFOs described in the following, are not fully optimized, but rather serve as a proof of principle for a relative performance enhancement. In section 2.4.3 the FFO has been introduced as a key element for integrated heterodyne high-frequency detectors. In order to be able to detect not only a high frequency, but also to do so with a fine resolution, the signal coming from the local oscillator (LO) in a heterodyne receiver needs to be both extremely stable and have a narrow bandwidth of the emitted frequency f_{LO} in order to achieve a high frequency resolution. From Eq. (2.61) it has already been established that this requires a stable bias voltage. In the case of a FFO as an LO, this means that the control-line current I_{cl} generating the flux penetration in the LJJ and the bias current I_{b} exerting a Lorentz force on the fluxons, need to be afflicted with as little fluctuation as possible.

For proper design of integrated receiver devices it is of great importance to know various design parameters of the employed fabrication process in detail. This includes the London penetration depth λ_{L}, the specific capacitance of the junction C^* and the parallel plate like structure of the wiring layer over the bottom electrode, *i.e.* the dielectric constant $\varepsilon_{\mathrm{ins}}$ of the insulation layer. Furthermore, the mutual inductance of the wiring layer over the bottom electrode, from which the rf matching circuit is typically patterned needs to be extracted. In section 3.3.2 the extraction of λ_{L} from interferometer measurements has been discussed as well as the extraction of the specific junction capacitance from the measurements of Fiske steps in *IVCs* [64].

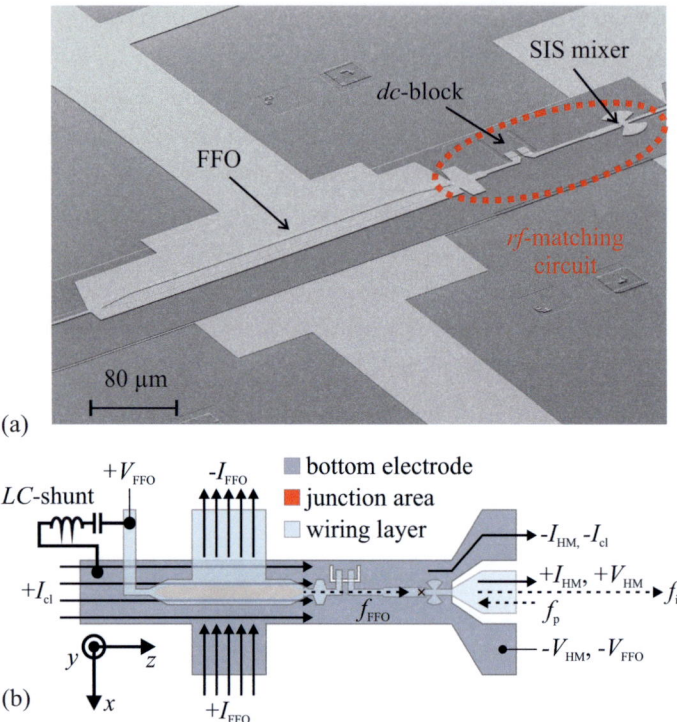

Figure 3.38.: In (a) a SEM image of the final device including the FFO, rf matching circuit and the harmonic mixer can be seen. The image was taken under a 60° angle. In (b) a schematic representation including all electrical connections is depicted. Solid arrows indicate dc currents, whereas dashed lines indicate rf signal paths.

ε_{ins} may be extracted from measurements of a coupled system between a multi-mode resonator and a short Josephson junction [156]. Essentially, the structure consists of a short JJ, strongly coupled to a superconducting microstrip resonator (SMR) formed from the Nb bottom electrode and Nb wiring layer. For a given structure, sub-harmonic pumping of the JJ will occur under the influence of an externally applied magnetic field H_{ext}, manifesting itself as current-steps in the sub-gap region of the JJ's IVC [156]. For structures fabricated in the conventional process, with an anodization voltage of $V_{\text{ao}} = 20\,\text{V}$ resulting in a Nb_2O_5 layer of $d_{\text{ao}} \approx 46\,\text{nm}$ and a subsequently evaporated SiO layer of $d_{\text{SiO}} \approx 300\,\text{nm}$ an effective dielectric constant of the insulation bilayer of $\varepsilon_{\text{ins}} = 9.5$ was extracted. Here, four different resonator geometries have been investigated with length $L_1 = 800\,\mu\text{m}$ and $L_2 = 1400\,\mu\text{m}$ and $W_1 = 80\,\mu\text{m}$ and $W_2 = 100\,\mu\text{m}$. The square junctions was always placed in the center of the resonator and had lateral dimensions of $W = L = 5\,\mu\text{m}$. For additional variation, critical-current densities of four different values $j_{\text{c}} = 0.8, 1.4, 2.0$ and $4.3\,\text{kA/cm}^2$ were chosen for fabrication.

From interferometer measurements the inductance per square of the wiring layer was

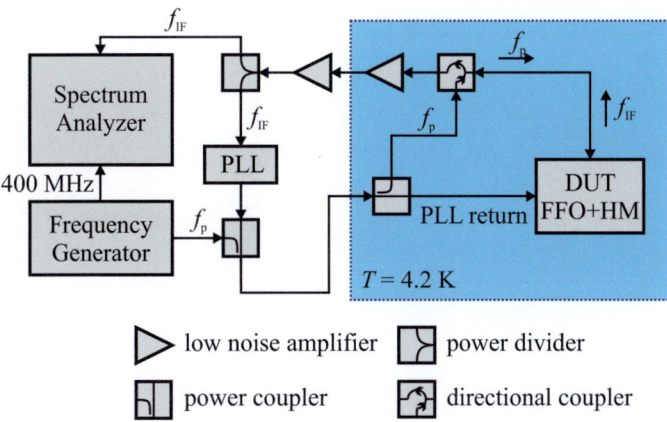

Figure 3.39.: Schematic of the measurement setup for a FFO linewidth measurement. The components kept at LHe-temperature are highlighted by a blue background.

extracted to be $L_{sq} = 660\,\text{fH}/\Box$, according to

$$L_{sq} = \mu_0 t_{ox,eff} \frac{L_w}{W_w}, \tag{3.18}$$

where μ_0 is the vacuum permeability and $t_{ox,eff}$ the effective barrier thickness. L_w and W_w denote the length and width of the inductance placed in the wiring layer. This value was extracted for a $12\,\mu\text{m}$ wide and $55\,\mu\text{m}$ long inductance in a $400\,\text{nm}$ thick Nb wiring layer. The critical-current density has been $j_c \approx 2\,\text{kA/cm}^2$.

Having extracted all required parameters, FFOs integrated on chip with harmonic mixers (HMs) were designed. A SEM image of such a device can be seen in Fig. 3.38 (a). The image was taken under a $60°$ angle. In the schematic representation depicted in (b) the paths and polarities of the dc bias currents for the FFO (I_{FFO}), the control line (I_{cl}) and the harmonic mixer (I_{HM}) are indicated in solid arrows. The ends of the FFO are tapered in order to suppress chaotic behavior [157]. Electro-magnetic radiation is coupled to the harmonic mixer via the rf matching circuit, which forms the rf connection between the FFO and the mixer. The dc block separates the biasing circuitry of the FFO from that of the mixer while still allowing progression of rf signals. The paths of the rf signals are indicated by dashed arrows. The LC-shunt is an optional external circuit, used in some measurements for the reduction of the emission linewidth and is not required for the principle operation of the device. The effect of the additional shunt will be addressed later in this section.

Fig. 3.39 depicts a block diagram of the rf connections of the measurement setup. The area highlighted in blue, indicates the parts inside the cryostat ($T = 4.2\,\text{K}$). All other parts were kept at room temperature without any intentional cooling. In order to be able to superimpose dc and rf signals, bias-Ts were used, separating the high frequency from the dc biasing. A frequency generator supplies a signal $f_p = k \cdot 20\,\text{GHz}$ via two power couplers

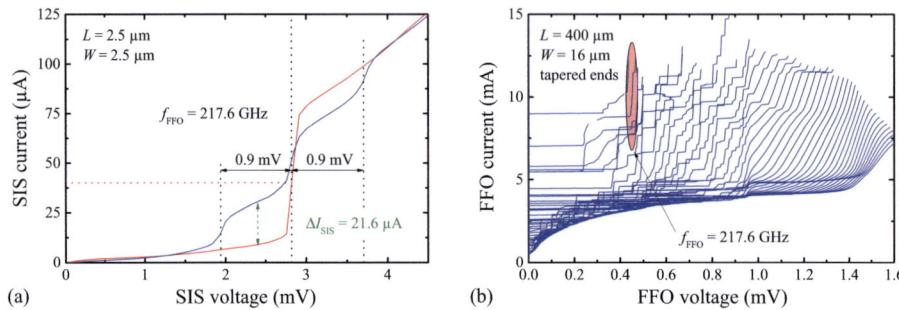

Figure 3.40.: (a) *IVC*s of a harmonic mixer with (blue trace) and without (red trace) LO-signal. The large current-step ΔI_{SIS} indicates a strong coupling of the FFO signal to the harmonic mixer. In (b) the corresponding *IV* family of the FFO is shown with the Fiske steps at $f_{FFO} = 217.6\,\text{GHz}$ highlighted.

and a directional coupler to the mixer. Thus the mixer is pumped with the kth harmonic f_p of a 20 GHz signal, coupled to the device using the identical co-planar waveguide as is used for the IF signal. The resulting IF frequency $f_{IF} = |f_{FFO} - k \cdot 20\,\text{GHz}| = |f_{FFO} - f_p|$ is therefore always below 20 GHz. Here, f_p is superposed with f_{FFO} coming from the on-chip local oscillator. The resulting IF signal f_{IF} is coupled via the directional coupler to a cold and subsequently a warm low noise amplifier (LNA) before it is split by a power divider. The IF signal is analyzed by a spectrum analyzer synchronized to 400 MHz coming from the frequency generator. For stabilization of the FFO, the IF signal is additionally fed to a frequency discriminator and fed back to the FFO, effectively creating a phase-locked loop (PLL) measurement mode. All rf signals were filtered at $T = 4.2\,\text{K}$ using an *LC*-filter, while the dc signals were filtered by a *RC*-filter. The shown measurement setup brings the advantage that only two coaxial cables had to be installed in the cryostat. Additionally, 5 twisted pair dc lines were installed for control of the harmonic mixer, the flux-flow oscillator and the control line. The differential transresistance $R_d^{cl} = \partial V_{FFO}/\partial I_{cl}$ of the control line with respect to the voltage across the FFO was extracted from I_{cl} and V_{FFO}.

The changes in the *IVC* of the harmonic mixer under the influence of a signal coming from the FFO can be seen in Fig. 3.40 (a). The red trace corresponds to the case without any FFO signal and the blue trace to one with a FFO signal. The equidistant current-steps around the gap voltage are clearly visible, indicating that a significant amount of LO-power is coupled to the mixer via the rf transmission line. In this particular case, the FFO has been emitting a frequency of $f_{FFO} = 217.6\,\text{GHz}$, resulting in current-steps at $\pm 0.9\,\text{mV} = f_{FFO} \cdot h/e$ away from V_g. The corresponding *IV* family of the FFO is depicted in Fig. 3.40 (b) with the Fiske steps at $f_{FFO} = 217.6\,\text{GHz}$ highlighted in red. The coupling efficiency from the FFO to the harmonic mixer, is typically determined by the height of the current-step at a fixed voltage in the *IVC* of the mixer. Here, the voltage was fixed to $V_p = 2.4\,\text{mV}$, resulting in a

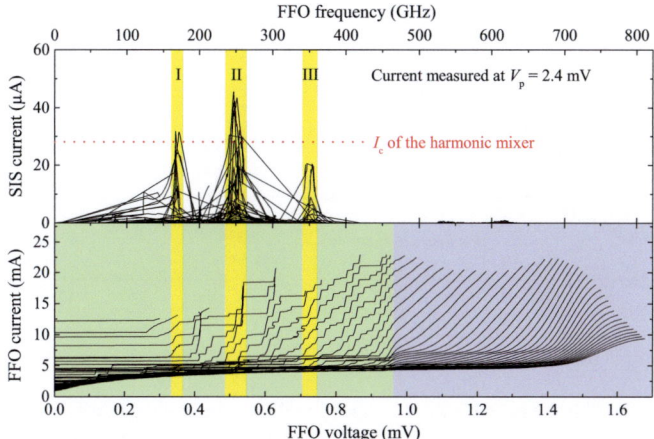

Figure 3.41.: Transmission behavior of rf matching circuit, measured by means of the pumping efficiency of the FFO to the mixer. The lower graph shows the *IV* family of a FFO for 51
different control-line currents. The Fiske regime is highlighted in green and the flux-
flow regime in blue. At the voltages highlighted in yellow, the emitted frequency is
transmitted via the rf matching circuit to the mixer. This results in an increased current
in the mixer, which is voltage biased at $V_p = 2.4\,\text{mV}$, depicted in black in the upper
graph. Additionally, the I_c of the mixer is indicated in red.

step height of $\Delta I_{SIS} = 21.6\,\mu\text{A}$, which is more than 50 % of the critical current $I_c \approx 40\,\mu\text{A}$
of the plain mixer. It should be noted, that the step height ΔI_{SIS} is typically taken to be the
total magnitude of the current at V_p, neglecting the sub-gap current.

For the investigation of the complete transmission behavior of the rf matching circuit,
the pumping efficiency as described in the previous paragraph, needs to be extracted over
a wide range of bias points of the FFO. In the lower part of Fig. 3.41 a family of *IVCs* of
a flux-flow oscillator is shown. Each trace corresponds to a different control-line current
and thus a different, locally applied magnetic field for the creation of fluxons in the LJJ.
The area highlighted in green corresponds to the Fiske regime. The current-steps are extremely steep and the voltage in fact jumps from each step to the next. Here, the LJJ is not
filled with fluxons and the steps result from a resonant excitement of the trapped vortices,
c.f. section 3.3.3. In consequence, the linewidth of the emitted frequency is very small as
compared to the linewidth in the flux-flow regime, which is highlighted in blue. In the flux-
flow regime, the LJJ is fully penetrated with Josephson vortices, resulting in a continuous
movement of the fluxons under the Lorentz force exerted by I_{FFO}. The resulting slope of
the steps is significantly shallower than in the Fiske regime, resulting in a larger linewidth
of the emitted frequency. However, this brings the advantage, that f_{FFO} can easily be tuned
continuously in the flux-flow regime, whereas tuning in the Fiske regime is rather complicated [100]. The horizontal feature at approximately $6 - 7\,\text{mA}$ results from pinned flux that
has been randomly trapped during cooling of the sample.

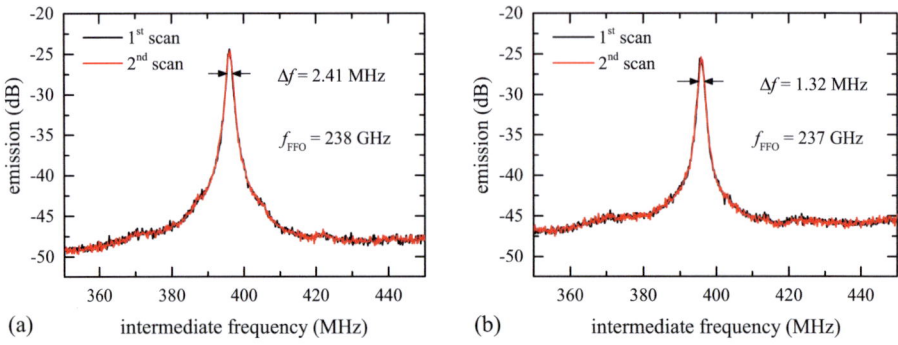

Figure 3.42.: Linewidth measurements of conventional (a) and an LC-shunted (b) FFO. Due to the negative feedback of fluctuations between the bias current and control-line current through the LC-shunt the linewidth is reduced.

In the upper part of Fig. 3.41 the current in the harmonic mixer, measured at $V_p = 2.4\,\text{mV}$ with respect to the voltage across the FFO is shown in black. The I_c of the mixer without any applied LO-power is indicated in red. The areas labeled with roman numerals I, II, and III, highlighted in yellow, correspond to the regions where strong coupling has been detected. The frequencies can be extracted from V_FFO using the second Josephson equation. From this data, it can be concluded that the rf matching circuit works for three frequency bands; I from 159.6 GHz to 174.1 GHz, II from 235.0 GHz to 264.5 GHz, III from 340.0 GHz to 360.3 GHz.

Based on considerations presented at the European Conference on Applied Superconductivity in 2009 [158], a new external circuitry for the enhancement of the FFO performance was investigated within this thesis. An extra LC-shunt, designed to have high-pass characteristics which do not interfere with the phase-locked loop mode of the FFO, as indicated in Fig. 3.38 (b) were connected in parallel to the FFO. As already discussed in section 2.4.3, such an LC-shunt can result in an artificial correlation of the FFO bias current I_FFO and the control-line current I_cl. In particular, the goal was to correlate the fluctuations of the two current sources, resulting in a cancellation and ultimately a reduction of noise influencing the FFO. By this, the frequency of the emitted radiation should be stabilized and reduced in linewidth. In Fig. 3.42 (a) and (b) two linewidth measurements of the identical FFO can be seen. (a) corresponds to the linewidth measurement without an extra LC-shunt and (b) to that with a shunt. Both measurements have been at comparable LO-frequencies $f_\text{FFO} \approx 237 - 238\,\text{GHz}$ and signal to noise ratios. As can be seen the $-3\,\text{dBm}$ linewidth is somewhat reduced when an additional serial LC-resonance circuit is connected in parallel to the FFO. This indicates that the fluctuations are indeed reduced due to a negative feedback through the shunt.

Conclusion

In this section it has been shown that the Nb/Al-AlO$_x$/Nb technology is suitable for fabrication of complex high-frequency Josephson devices such as flux-flow oscillators and SIS mixers. By means of interferometer measurements the mutual inductance of the Nb wiring layer over the Nb bottom electrode, as well as the London penetration depth was extracted [67, 68, 113]. Using a strongly coupled system of a short Josephson junction and a multi-mode resonator [156], the effective dielectric constant ε_{ins} of the multilayer of Nb$_2$O$_5$ and SiO was extracted. Finally, the specific junction capacitance was measured by measuring Fiske steps in a long Josephson junction. Using these process parameters, several designs for flux-flow oscillators, integrated on-chip with harmonic mixers have been developed. Final investigations have shown that an additional LC-shunt in parallel to the FFO, result in a correlation between the FFO bias current fluctuations and the fluctuations in the control-line current. Due to the resulting negative feedback, the total fluctuation can be suppressed to some degree, leading to a reduction of the frequency linewidth of the emitted radiation form the FFO.

4. Niobium Nitride and Aluminum Nitride Multilayers for Josephson Devices

In subsection 2.4.3 integrated receiver devices and in particular flux-flow oscillators were introduced as highly interesting applications for Josephson junction technology. As mentioned, the Nb/Al-AlO$_x$/Nb technology reaches an upper frequency limit at roughly $f_{\text{lim,Nb}} \approx 677\,\text{GHz}$, which strongly limits the ongoing research in the high-frequency spectrum. Attempting to master this challenge, the development of new multilayer fabrication technologies has been ongoing worldwide since the 1980's. Concerning the choice of electrode materials, most of these technologies are based on the superconductors niobium nitride (NbN), niobium titanium nitride (NbTiN) or combinations thereof with conventional niobium. Both NbN and NbTiN are typically deposited using reactive rf or dc sputtering from Nb targets or, in the case of NbTiN, in a compound procedure from Nb and Ti targets. Values of the energy gap of high-quality NbN found in literature range from $\Delta_{\text{NbN}} = 2.4 - 2.95\,\text{meV}$ [101, 159–162], which corresponds to a theoretical upper frequency limit of $f_{\text{lim,NbN}} \approx 2.85\,\text{THz}$ for the mixing element. For NbTiN typical values of the energy gap found in literature are also in the range of $\sim 2.5\,\text{meV}$ [163].

Concerning the desired increase in the critical-current density for achieving optimal ωRC-values, the conventional barrier material AlO$_x$ reaches a limit at around $j_c \approx 10\,\text{kA/cm}^2$, under the premise of $R_{sg}/R_N \geq 10$ [118, 119]. This can be understood easily, when looking at the potential barrier cooper pairs face upon tunneling from one electrode to the other. The coupling of the wavefunctions is defined by the strength of the decay within the barrier which strongly depends on the barrier thickness. With increasing band gap of the barrier material, the wavefunctions decay on a shorter length scale and consequently the critical-current density is reduced. From this it can be inferred, that an insulating material with a lower band gap than that of AlO$_x$, leads to higher values of j_c at equal thicknesses. In references [164, 165] an extensive overview of the different band structures of aluminum oxide and aluminum nitride is given. For bulk Al$_2$O$_3$ the band gap is determined to be $E_{\text{g,Al2O3}} = 8.8\,\text{eV}$, while a direct band gap of bulk AlN is found to be $E_{\text{g,AlN}} = 6.2\,\text{eV}$. With most of the junction technology being based on an oxidized aluminum barrier, the substitution of AlO$_x$ by AlN is often the obvious choice. For ease of fabrication in the case of NbN electrodes, the common substrate choice is magnesium oxide (MgO) in [100] orientation, since the lattice constant $a_{\text{MgO}} = 0.421\,\text{nm}$ fits that of NbN $a_{\text{NbN}} = 0.439\,\text{nm}$ [166] almost perfectly. Also, both materials exhibit a cubic lattice structure, hence promoting epitaxial growth even at low deposition temperatures.

First reports on NbN based Josephson junctions go back to the early 1980's [102–104]. Overall, there have been numerous publications of junctions based on NbN [101, 167–169]. So far the highest junction quality has been reported using a MgO tunneling barrier [162], exhibiting a gap voltage of up to $V_g = 5.9$ mV and a maximum critical-current density of $j_c = 70$ kA/cm^2. Using AlN as a tunneling barrier, the most noteworthy publications are [170–173]. Whereas most of the recent reports on NbN/AlN/NbN Josephson junctions are based on MgO substrates, only very few are based on silicon substrates [174, 175]. This is most likely due to the strong lattice mismatch between Si and NbN, resulting in poor initial growth of NbN. However, the use of Si as a substrate would have various advantages over MgO. Besides the apparent advantage of a significantly lower cost, Si is widely used in other technologies, such as RSFQ and SQUID technology. Thus, Si substrate would be beneficial for integration with such technologies. Furthermore, unlike MgO, Si is not hygroscopic. This plays a role, particularly for high frequency applications, where MgO becomes opaque due to absorbed water from the atmosphere, while high-resistive Si stays transparent. With back-illuminated detectors, this effect results in a significant difference in the radiation absorption and consequently detection efficiency.

Within this thesis, a technology for the fabrication of niobium nitride devices has been developed. This includes single layer NbN devices on different substrates including MgO, Al$_2$O$_3$ and Si, as well as multilayer devices of NbN in combination with aluminum nitride. In the following, the NbN/AlN/NbN technology will be described starting with the 3-chamber *in-situ* deposition system in the first section. The second section deals with the results of the optimization of niobium nitride films, while the third section addresses the integration of aluminum nitride as a buffer layer underneath the niobium nitride layer. The chapter closes with a discussion of NbN/AlN/NbN Josephson junctions on different substrates and the resulting characteristics.

4.1. Multi-Chamber Sputter System

During this thesis, a multi-chamber vacuum system for the *in-situ* deposition of epitaxial niobium nitride and aluminum nitride films for the fabrication of NbN/AlN/NbN Josephson junctions was designed and built. The main goal of this task, was to develop a process for all-NbN/AlN/NbN junctions based on 2″ silicon substrates, exhibiting good quality parameters for future applications in the THz frequency range.

Fig. 4.1 (a) shows an image of the 3-chamber system. The load-lock (top left) is equipped with an rf pre-cleaning system for the removal of native oxides on the silicon wafers. From the load-lock the substrate holder is moved to the NbN chamber (center) using a custom-built clamping mechanism. Fig. 4.1 (b) shows an image of the movable heater in the NbN chamber at $T = 775$ °C with the sample holder and samples on top. The resistive heater is powered by a current source [176], which is regulated using a PID controller [177]. Also

(a)

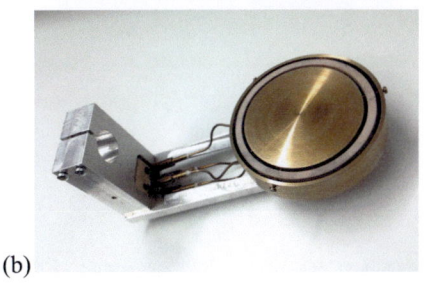
(b)

(c)

Figure 4.1.: (a) Image of the new 3-chamber deposition system, (b) image of the movable heater table (second generation) and (c) image of an ignited plasma inside the NbN chamber. The magnetron is visible at the top of the picture, the samples in the lower part on top of the heated table ($T_{dep} = 775\,°C$). The pivoting shutter can be seen at the very bottom of (c).

visible in (b) is the magnetron in the upper part and the pivoting shutter at the very bottom. In (c) an image of the second heater generation can be seen. The heater is mounted on a movable stage, allowing continuous heating while the sample is moved from the NbN chamber to the AlN chamber and back. Each of the three chambers is individually pumped by a turbo molecular pump and a rotary vane pump. For throttling of the gas flow in the NbN and AlN chamber, butterfly valves were modified in order to reduce the effective cross-section for the pumps. The base pressures range from $3 - 5 \times 10^{-5}\,Pa$. Both, the NbN as well as the AlN chamber are equipped with a $3''$ magnetron [178], powered by a $1\,kW$ sputter source [179].

Reactive sputtering systems leave a large parameter space for the deposition of thin films. Within this thesis, the properties of the niobium nitride films have been optimized with respect to the critical temperature T_c and the surface morphology, by means of varying

- the heater temperature during deposition T_{dep},

- the partial argon pressure p_{Ar},

- the partial nitrogen pressure p_{N2}, defining the percentage of the reactive gas and consequently also the required power density P applied to the target during deposition and

- the discharge current I_{dis}, defining the nitrogen consumption Δp_{N2} for a fixed set of the remaining parameters.

Additionally, aluminum nitride was introduced as a buffer layer underneath the NbN. As a buffer layer AlN enhances the lattice match between the Si[111] substrate and niobium nitride, promoting an improved crystallinity of NbN. Consequently, the critical temperature of NbN should be increased as compared to the T_c on pure Si substrates. Besides the increased T_c the AlN layer also works as an absolute etch stop for reactive-ion etching using a gas mixture of CF_4 and O_2 as is typically used for etching of Nb and NbN. For the use of aluminum nitride as a tunneling barrier it is of utmost importance to ensure homogeneous growth over the entire substrate without defects, which would create quasiparticle channels and therefore cause an increased leakage current of NbN/AlN/NbN JJs.

Conclusion

In this section the newly developed multi-chamber sputter system has been introduced. The system is equipped with an *in-situ* pre-cleaning in the load-lock and two 3″ magnetrons for the deposition of niobium nitride and aluminum nitride. For promotion of epitaxial growth on various substrates, a movable heater was installed, with a maximum operating temperature of $T \approx 850\,°C$. The system is suitable for the fabrication of high-quality epitaxial films of AlN and NbN on various substrates, forming an ideal basis for the development of all-NbN/AlN/NbN Josephson junctions. The optimization of the individual films will be introduced in the following.

4.2. Optimization of Niobium Nitride Films

For materials sputtered from a single target without the introduction of reactive components, the discharge characteristics are easily controlled through the applied power to the magnetron. In most cases, this is done by controlling the discharge current I_{dis} or the power applied to the target, resulting in a plasma discharge of the carrier gas. With a decreasing argon pressure, the amount of ionized argon is reduced, leading to a higher plasma impedance and thus discharge voltage V_{dis}, until a certain threshold is reached, below which, the applied power is not sufficient to sustain a stable plasma. This is also true for reactive sputter processes, however, depending on the amount of additional reactive gas N_2, the sputter conditions change drastically.

Fig. 4.2 shows three examples of current voltage discharge characteristics of reactive dc magnetron sputtering of niobium nitride. The argon pressures are 0.4 Pa, 0.6 Pa and 1 Pa for the three different traces; green, red and black, respectively. Solid squares correspond to sputtering of Nb in pure argon atmosphere and open circles show the behavior during the reactive process. The nitrogen content was varied, such that the *IV* characteristics are

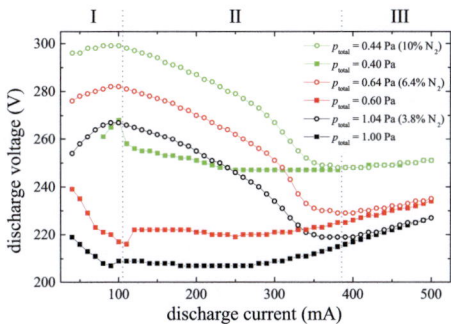

Figure 4.2.: Current voltage characteristics of reactive NbN sputtering for three different gas parameter sets. The solid squares correspond to sputtering of Nb in pure argon atmosphere.
Open circles show the discharge characteristics when nitrogen is added to the process.
Roman numerals I, II and III indicate the glow discharge regime, nitride phase and niobium phase, respectively.

comparable with respect to the discharge current. Three distinct regimes can be found in all
traces of the reactive processes, indicated with roman numerals (I-III).

In the first regime (I), the applied power is not sufficient for ignition of a stable plasma.
Phenomenologically, this results in a flickering of the plasma, fluctuations in I_{dis} and is
dominated by the so-called glow-discharge. Here, almost no material is sputtered from the
target, since the energy of the few ions is not sufficient to sputter particles from the target.
The sputter yield, which is defined to be the number of particle per primary ion [180], is
very low and approaches zero with decreasing discharge current.

The second regime (II) will be referred to as the nitrogen phase in the following. Here,
the plasma is stable, and the energy of the ions is sufficient to sputter etch the target. However, the nitrogen content in the gas mixture is such that it causes nitration of the target and
therefore an NbN compound is sputter etched from the target in this regime. Under these
conditions the target is considered to be *poisoned*. Although the sputter yield is much higher
than in regime I, it is still significantly lower than for pure niobium sputtered in argon atmosphere. Depending on the gas mixture, *i.e.* the partial pressure of the reactive component,
and the total gas-flow rate, a hysteresis as described in [181] may appear in the $p_{N2}(\mathscr{F}_{N2})$
characteristics during deposition, where $\mathscr{F}_{N2}$ denotes the nitrogen flow. Since the gas mixture is controlled by conventional needle valves instead of gas-flow meters, such characteristics as discussed in [181] cannot be extracted here. Nevertheless, a comparable effect can
be observed in the *IV* discharge characteristics for variation of the total flow at a constant
ratio between argon and nitrogen. With decreasing total flow, which may be controlled by
the throttling butterfly valve, the transition from regime II to III becomes increasingly steep,
until it becomes hysteretic. Despite the fact that such a hysteresis is unwanted for the deposition of NbN, for the sake of plasma homogeneity the total flow should not be increased far
beyond the point of its occurrence. For high nitrogen contents and high deposition powers,

small hystereses are very common and will be discussed later. With increasing discharge current the sputter yield increases further and more material is sputtered, until the sputter rate is equal and eventually larger than the nitration rate on the target. At this point the *IV* discharge characteristics approach the characteristic voltages for sputtering of Nb in pure argon atmosphere. In any case, the nitration takes place at the target, in the gas phase and on the substrate surface [182].

Beyond this point the curve enters the third regime (III) which will be referred to as the niobium phase. Here, the films consist of contaminated Nb of poor quality. Although the deposition rate of the material is almost identical to that of pure Nb, the quality parameters of the resulting NbN thin films are strongly degraded and typically exhibit critical temperatures well below that of pure niobium.

High-quality NbN films for use as ultra-thin film devices such as superconducting nanowire single-photon detectors (SNSPDs) or hot-electron bolometers (HEBs), as well as multilayer devices made of NbN/AlN/NbN trilayers, are deposited in the vicinity of the transition from the nitrogen phase to the niobium phase. The particular deposition parameters for the individual device depend on the required film thickness, surface morphology and in some cases also on the combination with layers of other materials.

It should be mentioned that the exact shape of the discharge characteristics strongly depends on various conditions, including the geometry of the sputter chamber and the surface of the target. Hence, a simple conversion of the deposition parameters from an existing system to a new system is not possible. Furthermore, fabrication of high-quality films requires continuous monitoring of all parameters, because the effective surface area of the target changes significantly over time [183]. Consequently the characteristics shown in this thesis may vary despite comparable parameters, due to the fact that 6 mm thick targets eroded over the course of this work. The graphs discussed in the following paragraphs may not display the global optimum of the deposition condition, but rather represent the specific steps leading to the final optimization. Nonetheless, the tendency of all shown graphs, represent the general behavior found for the optimization of this NbN/AlN/NbN technology.

In Fig. 4.3 (a) a change of the deposition temperature is shown. As can be seen, the maximum critical temperature depends on the heater temperature as well as the substrate choice. At each temperature two films have been deposited simultaneously on Si[111] and $Al_2O_3[1\bar{1}02]$ without any pre-cleaning procedure. Sapphire has been chosen as a reference substrate to Si[111], because at the IMS niobium nitride films on sapphire have been studied intensively over the last few years and the deposition parameters are well understood [94, 120, 184]. The values of T_c differ from $T_{c,bulk} \approx 16\,\mathrm{K}$, because the shown results are measurements on thin films that ranged from 8 nm to 10 nm in thickness. From (a) an optimal deposition temperature of $T_{dep} \approx 775\,°\mathrm{C}$ was extracted. Most of the following optimization steps on single and bi-layer films were done at this temperature. The second generation of substrate holders, as depicted in Fig. 4.1 (c), have a better thermal coupling to

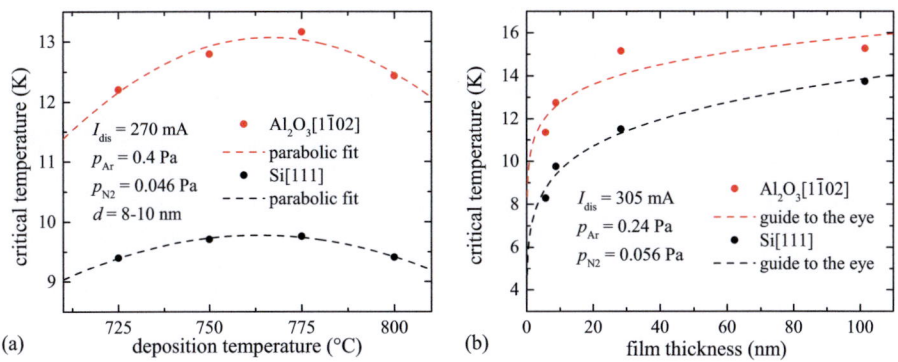

Figure 4.3.: (a) Shows the dependence of the critical temperature T_c on the deposition temperature of the heater and in (b) the T_c dependence on the film thickness is shown. Both curves correspond well with results found in other NbN deposition systems at the IMS.

the heater. Additionally, the total volume was decreased, resulting in an effective increase of the substrate temperature at identical heater temperatures. Consequently the optimal T_{dep} is lower for the final devices.

The thickness dependency shown in Fig 4.3 (b) behaves as expected from previously deposited films in other deposition systems at the IMS [185]. For optimization of the deposition atmosphere, it is therefore not necessary to deposit thick films as required for NbN/AlN/NbN JJs. In order to avoid the time consuming deposition of rather thick films at the order of $100 - 300\,nm$ at the low deposition rates in the nitrogen phase, thin films were deposited on Si[111] and $Al_2O_3[1\bar{1}02]$ instead. This made it possible to speed up the optimization procedure significantly.

From previous investigations at the IMS and several publications [161, 186] it is well-known, that the critical temperature of NbN tends to increase with a decreasing total deposition pressure. Of course this effect is limited by factors specific for each individual deposition system, such as the chamber geometry, the base pressure and the minimum pressure of the carrier gas at which the plasma is stable. In the case of the $3''$ magnetron used for the Nb target the lower limit for the plasma stability has been found to be $p_{Ar} = 0.2\,Pa$. However, films deposited at such low pressures exhibited strong compressive stress, to a degree where the films delaminated from the substrates once exposed to atmosphere. Fig. 4.4 shows two exemplarily SEM images of delaminated NbN films deposited at $p_{Ar} = 0.2\,Pa$ and $p_{N2} = 0.15\,Pa$. In (a) the entire film delaminated from the substrate, while in (b) the delamination occurred along hexagonally ordered paths. The pattern in (b) indicates that the films delaminate along the crystalline hexagonal surface structure of the Si[111] substrate. Already slightly increased argon pressures of $p_{Ar} = 0.24\,Pa$ did not show any signs of delamination, although probably still exhibiting compressive stress. For further optimization, $p_{Ar} = 0.24\,Pa$ was used.

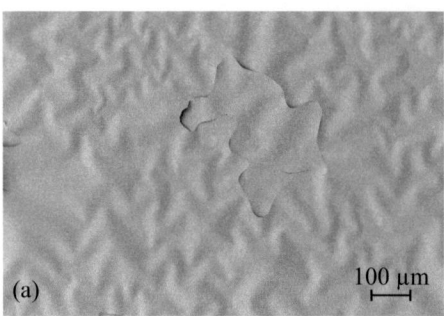 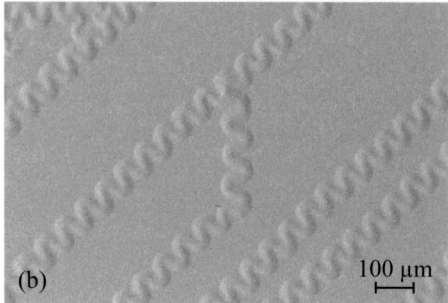

Figure 4.4.: NbN films deposited near the lower threshold of plasma stability at an argon pressure of $p_{Ar} = 0.2\,Pa$. In some cases the entire film delaminated (a) due to strong compressive stress, while few films delaminated along hexagonal paths (b).

Having found deposition parameters for comparably thin films below 30 nm with high critical temperatures, films of larger thicknesses were fabricated and investigated. In particular, the surface morphology and by extension the surface roughness had to be investigated and optimized for the use in multilayers for Josephson junctions. The key element for the fabrication of high-quality JJs is the homogeneity of the tunneling barrier. For Nb/Al-AlO$_x$/Nb JJs this challenge has been solved by the introduction of an aluminum wetting layer on top of the Nb bottom electrode [42, 43]. The Al levels the surface roughness of the Nb almost completely, allowing a homogeneous growth of thermal AlO$_x$, before the Nb counter electrode is deposited. While this approach is very promising for materials deposited at room temperature, it is unsuitable for heated depositions, since aluminum sputtered on heated substrates tends to form islands instead of wetting the surface of the bottom electrode. Unfortunately, NbN deposited on Si[111] substrate at room temperature will not yield a high T_c at a low enough surface roughness, making a heated deposition for NbN/AlN/NbN JJs on Si substrates necessary [187]. Consequently, the ideal fabrication process includes a heated deposition of a NbN/AlN/NbN trilayer, where the surface morphology and roughness of the bottom electrode allows the direct deposition of the extremely thin but defect-free tunneling barrier.

Despite a good uniformity and T_c of the so-far deposited thin films, when deposited for a longer time the films grew strongly polycrystalline, exhibiting an unacceptably large surface roughness. Fig. 4.5 shows detailed deposition characteristics for different fabricated samples. Fig. 4.5 (a) shows the *IV* discharge characteristics for pure argon atmosphere as black to gray solid squares. The shades correspond to the *IV* characteristics for gas mixtures of argon and nitrogen, depicted as solid circles. The partial argon pressure has been kept at $p_{Ar} = 0.24\,Pa$, while the partial nitrogen pressure has been varied form $p_{N2} = 0.057\,Pa$ to $p_{N2} = 0.149\,Pa$, *i.e.* approximately 20 % to 40 % nitrogen. As can be seen the hysteresis increases slightly with increasing nitrogen content, due to the rise in the total deposition pressure at equal pumping speed. Nevertheless, the resulting hystereses are small in mag-

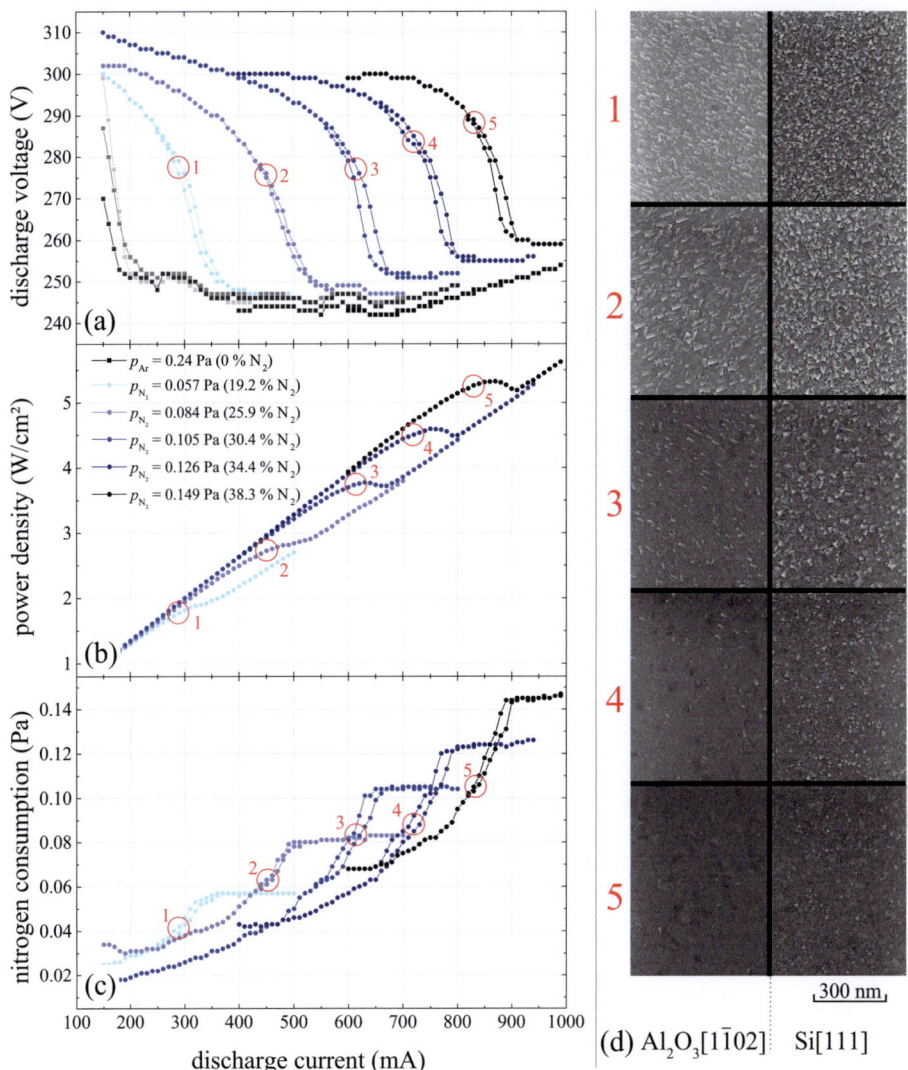

Figure 4.5.: Discharge characteristics (a), applied power density to the target (b) and nitrogen consumption (c) for different applied discharge currents. The different traces have been recorded for an argon pressure of 0.24 Pa and different nitrogen pressures. In (d) SEM images of films deposited at the deposition parameters indicated in (a-c) are shown.

Table 4.1.: Characteristic values of 10 NbN samples deposited under 5 different deposition parameters.

Sample (units)	P (W/cm^2)	Δp_{N2} (%)	d (nm)	$T_{c,Al2O3}$ (K)	$T_{c,Si[111]}$ (K)	RRR_{Al2O3} (1)	$RRR_{Si[111]}$ (1)
1	1.84	77.6	101	15.26	13.74	0.96	0.87
2	2.71	82.5	132	15.05	13.99	0.96	0.90
3	3.72	81.3	113	14.58	11.68	1.04	0.97
4	4.45	79.0	102	15.34	12.60	1.03	0.94
5	5.24	70.0	96	15.05	12.66	1.00	0.89

nitude and are mainly due to the speed at which the *IVCs* were measured. During actual deposition of the samples $1 - 5$, it was necessary to wait an appropriate time before opening the shutter, separating the sample from the plasma, in order to reach the equilibrium state. The resulting voltages are typically centered in the shown hystereses.

In Fig. 4.5 (b) the power-density dependences for the 5 different nitrogen contents are shown. The dc power is divided by the nominal target area of $A_{t,Nb} = 45.6\,\mathrm{cm}^2$. It should be noted that the traces were recorded after the niobium target had been in use for a significant amount of time. Thus, the shown values for the power density are over-estimates, since the effective target surface grows over time with increasing target erosion. Nevertheless, the power-density dependence may be mapped to the later investigation of the surface morphology.

The nitrogen consumption for the 5 different nitrogen contents is plotted in Fig. 4.5 (c). After [186] it is defined as the loss in partial nitrogen pressure upon ignition of a plasma,

$$\Delta p_{N2} = p_{Ar} + p_{N2} - p_{dis}. \tag{4.1}$$

Here p_{Ar} and p_{N2} denote the partial argon and nitrogen pressures respectively, and p_{dis} is the total pressure during stable plasma discharge. Whereas the pressure of the inert argon is unchanged upon plasma ignition, the nitrogen reacts with Nb, resulting in a power dependent consumption of the injected N_2. For small applied power, *i.e.* small discharge currents, the consumption tends towards a global minimum, while it continuously rises throughout the nitrogen phase with increasing I_{dis} until it eventually reaches $\Delta p_{N2} \lesssim 100\%$ when entering the niobium phase. With an increasingly steep transition from the nitrogen phase to the niobium phase, the nitrogen consumption also becomes steeper, making a fine tuning extremely complicated. This effect is also reflected in the occurrence of the negative trend of the $P(I_{dis})$ dependence for large values of p_{N2} at the transition from regime II to III, as shown in (b).

At each of the indicated deposition parameters (1-5), two samples were deposited si-

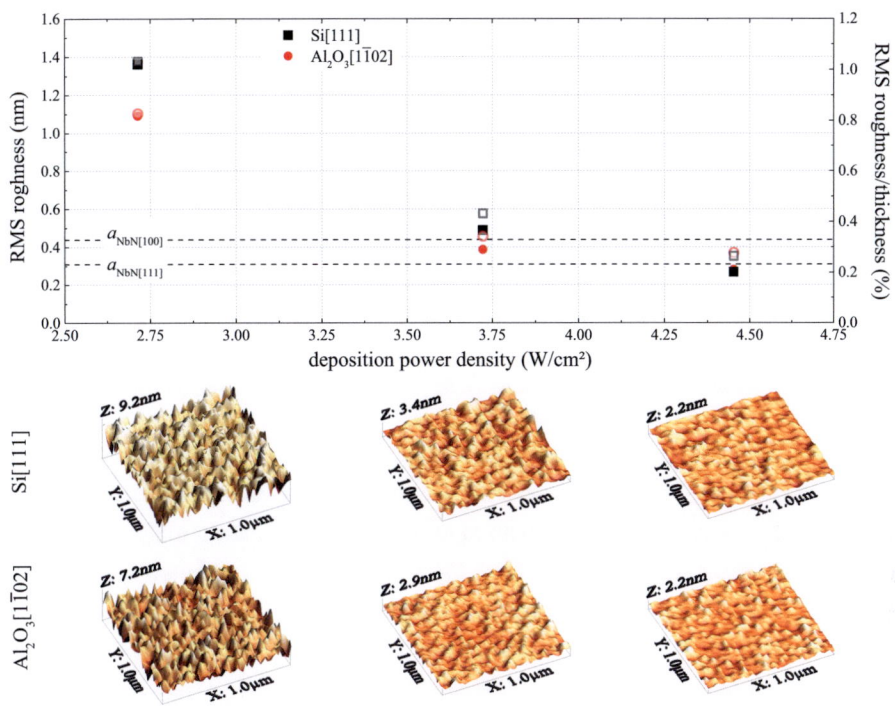

Figure 4.6.: Atomic force microscope scans of NbN films deposited at different power densities. A clear trend towards smoother surfaces can be seen with increasing sputter power density. The root mean square (RMS) surface roughness over a $1 \times 1\,\mu m^2$ scan field, even drops below the dimension of a single unit cell of NbN for $P \gtrsim 4\,W/cm^2$.

multaneously at $T_{dep} = 775\,°C$ on $[1\bar{1}02]$-oriented Al_2O_3 and [111]-oriented silicon. In Fig. 4.5 (d) SEM images of the 10 sample surfaces are shown. Care has been taken to deposit the samples under comparable conditions, such that comparable NbN thicknesses were reached. The characteristics are summarized in Tab. 4.1. The variance in the characteristic features are insignificant for the investigated surface structure. From the SEM images it can be inferred that the smoothness increases with increasing deposition power density, indicating that for thick NbN layers, as used in NbN/AlN/NbN JJs, high power densities are required in order to achieve flat surfaces. For thin layers below 30 nm, the power density on target plays only a minor role, since the crystalline structure is still dominated by the initial growth on substrate, rather than the inherent crystalline arrangement in thick layers.

The surface roughness was investigated in greater detail for samples 2, 3 and 4 at the *Physikalisches Institut*, KIT, Karlsruhe, Germany. From the SEM images shown in Fig. 4.5 (d) a transition from rough to smooth surfaces was expected. Using an atomic force microscope areas of $1 \times 1\,\mu m^2$ were scanned and the root mean square (RMS) and peak to peak values of the surface roughness were extracted. Fig. 4.6 depicts a plot of the RMS

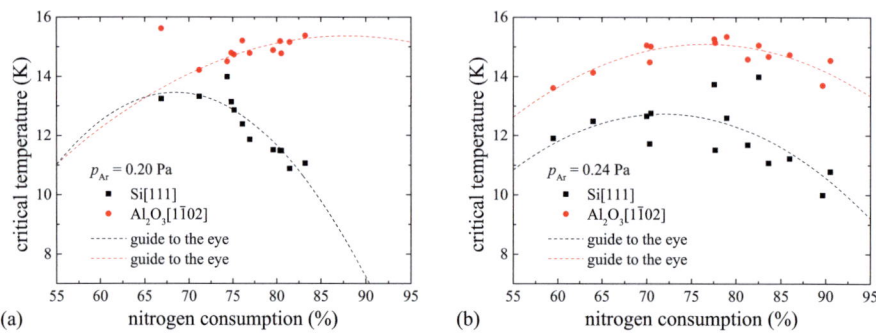

Figure 4.7.: Dependence of the critical temperature on the nitrogen consumption for an argon pressure of (a) 0.20 Pa and (b) 0.24 Pa.

surface roughness over the power density applied to the target (solid symbols). Additionally the RMS surface roughness was normalized to the total film thickness to eliminate effects of reduced roughness due to thinner films (open symbols). As can be seen, both values decrease with increasing power density. It should be noted, that this effect contradicts the commonly accepted assumption that the surface roughness increases with increasing deposition rate for niobium nitride as the deposition rate scales with the power density. The RMS roughness even drops below the lattice constant of NbN[100] and NbN[111]. Despite the striking difference between the SEM images of NbN on $Al_2O_3[1\bar{1}02]$ and on Si[111] the roughnesses measured using an atomic force microscope do not differ significantly for high power densities and for values of $P \gtrsim 4\,W/cm^2$ are almost identical. 3D images of the atomic force microscope scans are shown as well in Fig. 4.6. The z-scales have been adjusted to the largest measured peak to peak value of all six scans. Again it becomes obvious that there is a clear trend towards lower peak to peak values with increasing power densities.

From Eq. (2.4) it is known that the critical temperature is an implicit measure for the energy gap of the superconductor. In order to fabricate Josephson junctions suitable for high frequency applications, a large gap voltage is required, as has been derived in subsection 2.4.3. Having found a way to fabricate smooth NbN samples even at large thicknesses, optimization of the critical temperature of such samples became necessary. In various publications [161, 181, 188, 189] it has been shown that the flow of the reactive gas is of utmost importance for controlling the stoichiometry and thus characteristic parameters of the resulting films. Here, the approach originally introduced by Thakoor *et al.* [186] has been adapted, investigating the dependence of the critical temperature on the nitrogen consumption. Thakoor *et al.* found that the optimal T_c is found close to the point where $\Delta p_{N2}(p_{N2})$ starts deviating from a linear dependence. Whereas in [186] the nitrogen injections was varied by changing the N_2 flow, here the nitrogen consumption has been varied by means of the discharge current. This approach allows extremely fine tuning of Δp_{N2} in

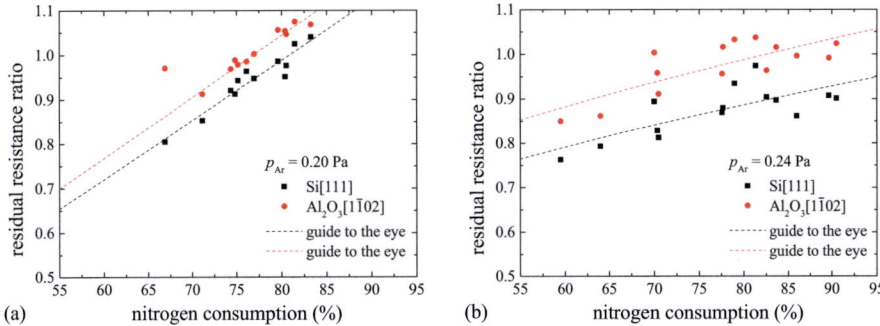

Figure 4.8.: Dependence of the residual resistance ratio on the nitrogen consumption for an argon pressure of (a) 0.20 Pa and (b) 0.24 Pa.

the case of non-hysteretic *IV* discharge characteristics. Fig. 4.7 shows the dependences of the critical temperature on the percental nitrogen consumption for a carrier gas pressure of $p_{Ar} = 0.20\,\text{Pa}$ (a) and $p_{Ar} = 0.24\,\text{Pa}$ (b). The thickness of all samples is larger than 100 nm, so that thickness related suppression of T_c can be neglected.

Similarly to the results from Thakoor *et al.* [186], the optimal T_c was found near the maximum nitrogen consumption, coinciding with the transition region from the nitrogen phase to the niobium phase. However, thorough investigation revealed that the optimal point is not exactly at the onset of maximum consumption but rather depends on the argon pressure, or flow, and furthermore may be strongly dependent on the substrate choice. Where argon pressures are very low (a), there exists a striking difference in the optimal percental N_2-consumption for sapphire and silicon substrate, whereas for even slightly higher argon pressures (b) the optima seem to be located near comparable Δp_{N2}. This trend continues for even higher argon pressures, resulting in almost the same optimal Δp_{N2} for $p_{Ar} = 0.40\,\text{Pa}$. For constant argon pressures and pumping speed, the percental $\Delta p_{N2}(T_{c,max})$ was found to be independent of the nitrogen injection p_{N2} in the case of non-hysteretic *IVC*s. For all samples, the critical temperature on sapphire substrate is found to be noticeably higher than on silicon substrate, although there might occur a crossing in the case of $p_{Ar} = 0.20\,\text{Pa}$ far in the nitrogen phase. Due to the low T_c-values in this region, this has not been investigated further. In the case of $p_{Ar} = 0.24\,\text{Pa}$, there seems to be an almost constant offset over the entire investigated Δp_{N2}-range. This difference may be ascribed to the worse crystalline structure of the NbN films on Si[111] than on $Al_2O_3[1\bar{1}02]$ as shown in Fig. 4.5 (d).

Next to the $T_c(\Delta p_{N2})$ dependence, the residual resistance ratio, RRR, was also investigated with respect to the nitrogen consumption. For better comparability, here the RRR is defined as the ratio between the room temperature resistance R_{300} and the resistance R_{20} at $T = 20\,\text{K}$. In Fig. 4.8 the RRR values are plotted over Δp_{N2} for $p_{Ar} = 0.20\,\text{Pa}$ in (a) and $p_{Ar} = 0.24\,\text{Pa}$ in (b). Again, the samples deposited on sapphire exhibit higher quality, due

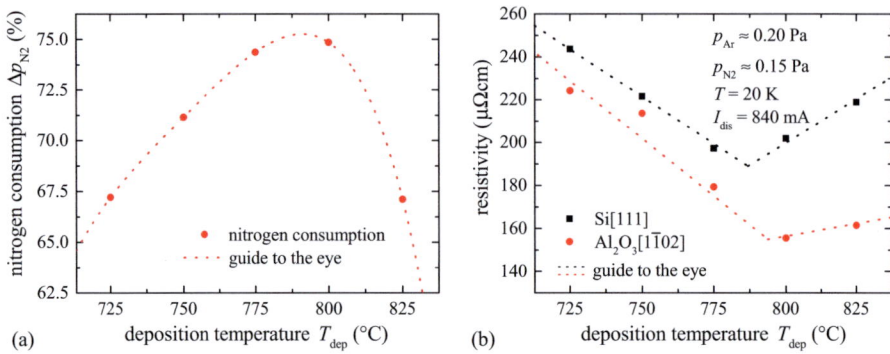

(a) (b)

Figure 4.9.: The nitrogen consumption (a) and the resistivity (b) for different deposition temperatures. All other sputter parameters have been kept constant for this investigation and are summarized in figure (b).

to better crystalline growth than on silicon substrates. The dashed lines are merely guides to the eye, as the RRR levels upon reaching the niobium phase at $\Delta p_{N2} \to 100\,\%$.

In the case of heated deposition of the NbN films, the nitrogen consumption is not only defined by the power applied to the target, but also by the power applied to the heater. Due to an increased surface energy, heating of the substrates promotes nitration on the sample surface. Fig. 4.9 (a) shows the dependence of the nitrogen consumption on the deposition temperatures. The shown depositions have all been performed at $p_{Ar} \approx 0.20\,\text{Pa}$, $p_{N2} \approx 0.15\,\text{Pa}$ and $I_{dis} = 840\,\text{mA}$. The deposition temperature at which the nitrogen consumption reaches a maximum, coincides with the previously found optimal T_{dep} with respect to the maximum T_c. This indicates that for higher temperatures nitrogen cannot be absorbed by the Nb, resulting in a decrease of the nitrogen consumption for further increased deposition temperatures.

This behavior is directly reflected by the quality parameters of the deposited films. In Fig. 4.9 (b) the resistivity of the films at $T = 20\,\text{K}$ is plotted over varying deposition temperatures. Up to the optimal temperature, the resistivity decreases, which corresponds to a higher film quality and also to an increase in the critical temperature. For higher temperatures the resistivity rises, indicating a reduced film quality. The reason for this behavior has been found in other deposition systems at the IMS as well, although at slightly different temperatures for silicon. In the case of sapphire, this effect has not been observed so far. In any case, from these measurements it can be concluded that deposition of NbN films on silicon substrates for NbN/AlN/NbN JJs needs to be done below a deposition temperature of $T_{dep} = 800\,^{\circ}\text{C}$.

In thin granular, or epitaxial, films of reasonably good quality, the London penetration depth $\lambda_{L,NbN}$ is typically in the range of $\sim 180 - 550\,\text{nm}$ [190–194], with the single exception of $\lambda_{L,NbN} \approx 90\,\text{nm}$ in [195]. Due to this typically significantly larger London penetra-

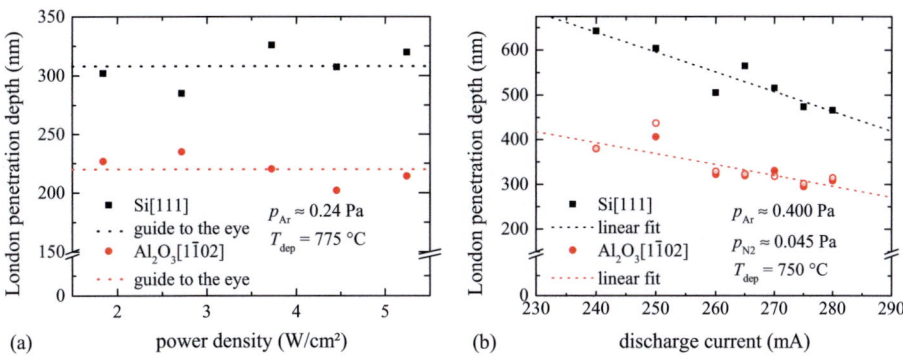

Figure 4.10.: London penetration depths calculated according to Eq. (4.2) [197]. (a) Shows the calculated values for samples fabricated at different deposition parameters, whereas (b)
shows the $\lambda_L^{GL}(I_{dis})$ dependence for one specific set of deposition parameters as indicated in the graph. Open transparent symbols correspond to R_N-values extracted from
IVC measurements at $T = 4.2\,\text{K}$.

tion depth of NbN as compared to Nb ($\lambda_{L,Nb} \approx 90\,nm$), it is necessary to fabricate JJs with
relatively thick electrodes $d_{electrodes} \approx \lambda_{L,NbN}$, such that any suppression of the effective gap
voltage due to suppressed energy gaps in the electrodes can be ruled out [196]. Never-
theless, for ease of patterning, the electrode thickness should be kept at a minimum, thus
minimizing the negative effects regarding the isotropic etching process and the resulting
overetching during patterning of the trilayer. Fig. 4.10 (a) shows calculated values of the
London penetration depth $\lambda_{L,d}^{GL}$ in the dirty limit

$$\lambda_{L,d}^{GL} = 64.2 \cdot 10^{-9} \sqrt{\frac{\rho}{T_c}} \cdot \left(1 - \frac{T}{T_c}\right)^{-1/2} \tag{4.2}$$

for the 10 samples already discussed earlier. Here, the resistivity ρ is in units of $\mu\Omega\text{cm}$.
Eq. (4.2) is an estimation based on the Ginzburg-Landau theory [26] derived by Orlando
et al. [197]. The measurements in Fig. 4.10 (a) are based on rather crude estimations of the
film's cross-section, but they do however show that the difference in the magnetic penetration depth between the films deposited on different substrates is independent of the applied
sputter power. It should be noted that the values were calculated under the assumption that
the normal state resistance at $T = 4.2\,\text{K}$ from IVC measurements is approximately equal to
the resistance of the film at $T = 20\,\text{K}$ from $R(T)$ measurements.

The transparent open symbols shown in Fig. 4.10 (b) are based on much more precise
values of the NbN-resistivity on $Al_2O_3[1\bar{1}02]$ substrate. These were extracted from IVC
measurements at $T = 4.2\,\text{K}$ on bridges of 4 squares. The solid symbols correspond to values
based on $R(T)$ measurements. The coincidence of the penetration depths calculated from
$R_{N,4.2K}$ and R_{20K}, validate the above made assumption. For the shown samples a given set of

gas mixture parameters was employed, while the deposition current was varied for each run. In order to achieve a pronounced $\lambda_L^{GL}(I_{dis})$ characteristic, thin films with comparably large resistivities were investigated. For films of larger thicknesses this dependence becomes less pronounced, however follows the same trend as shown in here. Again, the optimal condition with respect of the penetration depth is found when approaching the niobium phase. Particular care needs to be taken not to deposited degraded Nb by applying too much power, since $\lambda_L^{GL}(I_{dis})$ follows a comparable behavior as the resistivity shown in Fig. 4.9 (b).

Conclusion

From the above described investigations it can be concluded that the new deposition system is suitable for fabrication of high-quality NbN films. Proper deposition parameters were found after thorough investigation of a large parameter space. Initially, the optimal partial argon pressure was investigated and found to be $p_{Ar} = 0.24\,\mathrm{Pa}$, resulting in high critical temperatures at acceptable film stress. The optimal deposition temperature, with respect to the high critical temperature, low resistivity and low magnetic penetration depth has been found to be $T_{dep} \lesssim 775\,^\circ\mathrm{C}$. For values of $T_{dep} \gtrsim 800\,^\circ\mathrm{C}$ a strong degradation of the film quality was observed. It should be noted that the exact temperature may be specific to the geometry of each individual deposition system, however this effect has been observed similarly in other available systems.

Based on the previous investigation of Thakoor *et al.* [186], here the nitrogen consumption has been investigated in greater detail. It has been found that the exact optimum of Δp_{N2} with respect to a maximum T_c depends on the sputter atmosphere and simultaneously on the choice of substrate. Consequently, the percental nitrogen consumption has been identified as a global indicator for the quality of the deposited NbN films. For the optimal argon pressure, the highest T_c was found at $\Delta p_{N2} \approx 75\,\%$. Fine tuning of the nitrogen consumption may be done by means of varying the deposition temperature and discharge current. This allows almost independent variation of the partial nitrogen pressure p_{N2} and thus variation of the power applied to the target, which in turn has been identified as the prime parameter affecting the surface morphology. Power densities of $P \gtrsim 4\,\mathrm{W/cm^2}$ resulted in acceptably smooth surfaces of NbN layers on various substrates even for film thicknesses above 200 nm. Applying all these studies to the deposition of NbN allows precise tuning of the NbN quality parameters for a large variety of applications, including NbN/AlN/NbN JJs, HEBs and SNSPDs.

4.3. Integration of Aluminum Nitride Films for Niobium Nitride Devices

This section focuses on the integration of aluminum nitride films for various applications based on niobium nitride. In subsection 2.4.3 the advantage of silicon substrates over other materials such as sapphire or magnesium oxide was addressed. Unfortunately, patterning of NbN films on silicon is complicated, since there is no method available for selective etching. Using physical etching like IBE damages the substrate because the etching rate of Si is somewhat higher than for NbN. In the case of reactive-ion etching the effect is even stronger, resulting in an enormous overetching since Si is etched significantly stronger than NbN using the available RIE system. Particularly for patterning of thick films, an etch stop layer is desirable.

Additionally, the native oxide on silicon can result in damage to the first few atomic layers of NbN ultra-thin films and thus reduce the effective thickness, resulting in decreased quality parameters. The lattice mismatch between silicon and niobium nitride further degrades the films. Whereas diamond cubic silicon has a lattice of $a_{Si} = 0.543$ nm, niobium nitride has a face centered cubic lattice with $a_{NbN} = 0.439$ nm. This strong mismatch of $m_{NbN[100]\text{-}Si[100]} = 23.69\%$ leads to poor initial growth of NbN on Si. Here the mismatch between two lattices A and B is calculated by:

$$m_{\text{A-B}} = \left| \frac{a_A - a_B}{a_A} \right|, \tag{4.3}$$

where $a_{A,B}$ denotes the lattice spacing for a given orientation. For an overview, Tab. 4.2 summarizes the lattice constants a and in-plane spacings of materials typically used for different orientations. It is obvious that the high NbN film quality on MgO[100] is related to the low mismatch $m_{NbN[100]\text{-}MgO[100]} = 4.33\%$.

In order to circumvent the strong mismatch between NbN[100] and Si[100], differently oriented silicon substrate and the use of an aluminum nitride buffer layer has been reported by Shiino et al. [166]. Despite the rather large mismatch between Si[111], having a hexagonal surface structure and AlN[0001] of $m_{AlN[0001]\text{-}Si[111]} = 23.47\%$, epitaxial growth of

Table 4.2.: Overview over lattice constants and d-spacings for materials typically used for NbN based devices. Values were taken from [166, 198–203]. For Si, MgO and NbN Miller indices and for Al_2O_3 and AlN Bravais indices are given.

Material	Surface structure	a (nm)	c (nm)
Si[100]	diamond cubic	0.543	-
Si[111]	hexagonal	0.384	-
$Al_2O_3[0001]$	wurtzite hexagonal	0.476	1.299
$Al_2O_3[1\bar{1}02]$	tetragonal	0.476	1.539
MgO[100]	face centered cubic	0.420	-
MgO[111]	hexagonal	0.297	-
AlN[0001]	wurtzite hexagonal	0.311	0.498
NbN[100]	face centered cubic	0.439	-
NbN[111]	hexagonal	0.310	-

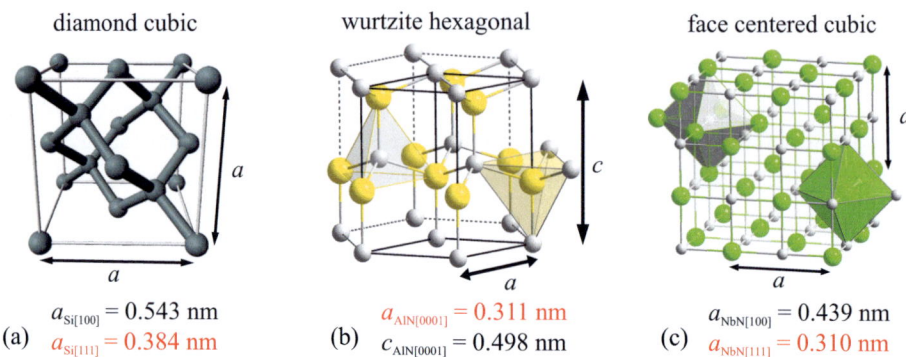

diamond cubic   wurtzite hexagonal   face centered cubic

(a) $a_{Si[100]} = 0.543$ nm
$a_{Si[111]} = 0.384$ nm

(b) $a_{AlN[0001]} = 0.311$ nm
$c_{AlN[0001]} = 0.498$ nm

(c) $a_{NbN[100]} = 0.439$ nm
$a_{NbN[111]} = 0.310$ nm

Figure 4.11.: Schematic representation of (a) diamond cubic, (b) wurtzite hexagonal and (c) face centered cubic lattices structures. The lattice constants in the optimal orientation for silicon - aluminum nitride - niobium nitride multilayers are highlighted in red. The images were adapted from [204].

AlN can still be achieved along the parallel oriented Si[2$\bar{2}$0] and AlN[11$\bar{2}$0] planes with interlayer dislocations such that $m_{5AlN[11\bar{2}0]-4Si[2\bar{2}0]} = 1.01\%$ [205–208]. In this scenario the growth direction of AlN remains to be the [0001] orientation parallel to Si[111] direction. It should be noted that throughout this work, Miller indices are used for materials with isotropic lattices like Si, MgO and NbN, whereas Bravais indices are given for materials with anisotropic lattices, such as Al$_2$O$_3$ and AlN [209].

Under the assumption of an epitaxial AlN buffer layer it should be possible to deposit hexagonal oriented NbN[111] epitaxially, since the mismatch $m_{NbN[111]-AlN[0001]} = 0.32\%$ is almost non-existent. Additionally, this process brings the advantage of avoiding the interface between NbN and the native oxide of the substrate. The corresponding crystalline structures for silicon (diamond cubic), aluminum nitride (wurtzite hexagonal) and niobium nitride (face centered cubic) are depicted in Fig. 4.11 (a-c), respectively. The relevant lattice constants of the orientations with the least mismatch are highlighted in red.

In order to be able to grow AlN[0001] on Si[111] epitaxially, removal of the native oxide is required since the native SiO$_x$ has an amorphous structure. Without a clean interface between the two hexagonal structures of Si[111] and AlN[0001], lattice-matched deposition is almost impossible. A common approach for the removal of native oxides is dipping the sample in diluted hydrofluoric (HF) acid [207, 208]. This procedure does however require a fast evacuation of the load-lock chamber, such that the sample does not oxidize again after the HF-dip. With a large load-lock chamber, as installed in the new deposition system, a sufficiently fast evacuation cannot be ensured. For the oxide removal, rf etching was therefore installed, allowing *in-situ* cleaning of the surface. The physical etching is typically performed at low power $P_{rf} = 20$ W and high Ar-pressure $p_{Ar} = 10$ Pa, resulting in a very low etching rate $er_{pc} = 0.01$ nm/s for smooth surfaces. From ellipsometric measurements, the typical thickness of the native oxide on high resistive Si[111] wafers was found to be

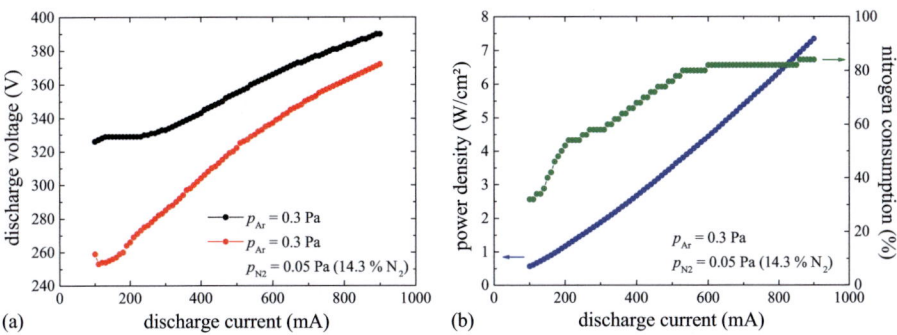

Figure 4.12.: The *IV* discharge characteristics of an aluminum nitride deposition is shown in (a). Due to strong emission of secondary electrons, the voltage drops upon injection of N_2. In (b) the power-density dependence (blue trace) and the nitrogen consumption (green trace) is depicted.

$d_{\text{SiO,nat}} \approx 0.9\,\text{nm}$ and will thus be removed completely during a 10 min etching procedure.

Aluminum nitride deposited on an *in-situ* cleaned silicon substrate prevents diffusion of nitrogen from the NbN film to the substrate. Thus, after rf cleaning, the load-lock is immediately evacuated to $p_{\text{LL}} < 2 \cdot 10^{-4}\,\text{Pa}$, before the sample is then transferred to the NbN chamber. Upon transition to the AlN chamber, the heating process is started. AlN is then deposited in a mixture of argon and nitrogen. Unlike for NbN deposition, the discharge voltage drops with the introduction of nitrogen, which is shown in Fig. 4.12 (a). This effect is addressed in detail by Lewis *et al.* [210], who concluded that the discharge behavior for AlN in an argon-nitrogen mixture must be dominated by the secondary electron emission rather than the plasma density. With the increased amount of free electrons, the discharge voltage decreases. For sputtering of niobium in an argon-nitrogen atmosphere the discharge characteristics are dominated by the plasma density leading to an increase of the discharge voltage.

Fig. 4.12 (b) shows the power-density dependence and the percental nitrogen consumption over varying discharge currents as blue and green traces respectively. Due to a high pumping speed the $P(I_{\text{dis}})$ dependence is almost linear and exhibits no hysteresis. This results in very stable process parameters. Additionally, the fact that during deposition the highly insulating compound AlN is formed prevents the formation of shortcuts between the cathode and anode, making it a process of extraordinary long-term stability. The high pumping speed also leads to a saturation effect of the maximum nitrogen consumption at $\Delta p_{\text{N2}} \lesssim 80\%$.

Similarly to the nitrogen consumption during reactive sputtering of niobium nitride, Δp_{N2} also exhibits an almost linear rise with increasing nitrogen injection p_{N2} for the reactive sputtering of aluminum nitride. However, the $\Delta p_{\text{N2}}(p_{\text{N2}})$ dependence exhibits a more pronounced and sudden drop for AlN than for NbN, once the target is poisoned. In general

117

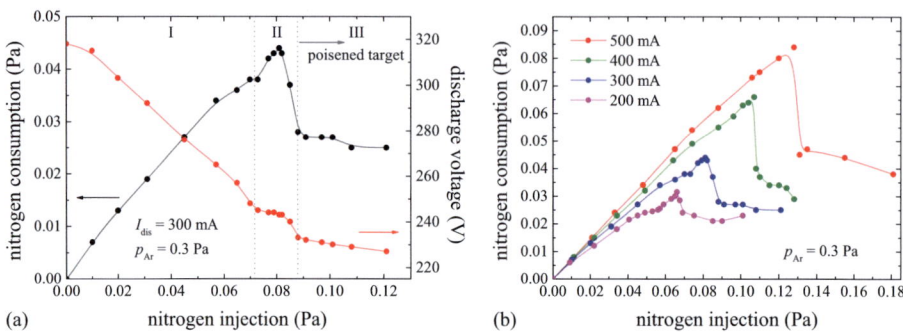

Figure 4.13.: In (a) the nitrogen consumption Δp_{N2} (black trace) and discharge voltage V_{dis} (red trace) is shown with respect to the nitrogen injection p_{N2}. In (b) the $\Delta p_{N2}(p_{N2})$ dependence for different discharge currents is depicted. The solid lines are guides to the eye.

the $\Delta p_{N2}(p_{N2})$ curve in Fig. 4.13 (a) can be divided in three different parts (I-III) [207]. In the first region the injected nitrogen is fully absorbed by the sputtered aluminum, *i.e.* the sputter rate is larger than the nitration rate. Here, the discharge voltage continuously drops according to the increase in the secondary emission of electrons [210]. In region II the target is partially nitrided, resulting in a deviation of the almost linear $\Delta p_{N2}(p_{N2})$ dependence in region I, and a very shallow drop in the voltage. In this regime the sputter and nitration rate are almost identical. Once the target is fully nitrided, or poisoned, the nitrogen consumption and discharge voltage V_{dis} exhibit a sudden drop; the curve enters region III. With increasing nitrogen injection, both Δp_{N2} and V_{dis} remain at an almost constant value, depending on the exact trade-off between the increase in the secondary emission of electrons and the decrease of the ionization cross-section [211].

In Fig. 4.13 (b) the nitrogen consumption is shown for various different discharge currents over the nitrogen injection. Here, the partial argon pressure is kept constant at $p_{Ar} = 0.3\,\text{Pa}$ for all characteristics. A comparable behavior to that described above can be seen for all curves. As expected, the required nitrogen partial pressure for poisoning of the target increases with increased discharge current, as the primary sputter rate is proportional to the power applied to the target. In all cases, the transition from metallic to semiconducting behavior takes place in region I. In the quite narrow region II the characteristics of the deposited film rapidly vary from semiconducting to insulating, resulting in strongly insulating properties in region III. Here the typical resistivity is at the order of $\rho_{AlN} \approx 10^{12}\,\Omega\text{cm}$. Also, the films become transparent, with a refractive index extracted from ellipsometric measurements at $\lambda_{opt} = 635\,\text{nm}$ of $n \approx 2$, coinciding well with values from literature [212]. It should be noted that despite the almost constant discharge voltage and current in region III, the deposition rate continuously drops with increasing nitrogen injection.

To estimate an optimal discharge current of AlN with respect to the critical temperature of NbN, four films of approximately 10 nm AlN were deposited on Si[111]. All substrates

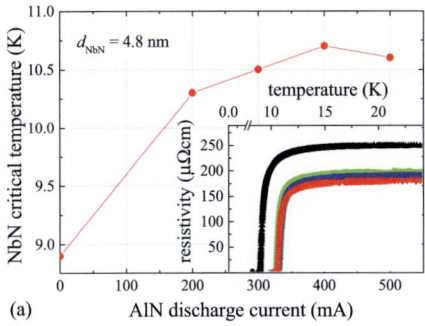 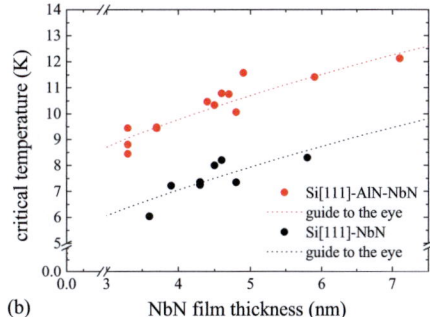

(a)   (b)

Figure 4.14.: (a) Shows the critical temperatures of 5 NbN film simultaneously deposited on differently prepared Si[111]-AlN substrates. The sample indicated at an AlN discharge current of zero corresponds to a plain Si[111] substrate. The inset shows the corresponding $R(T)$- measurements. In (b) the dependence of the critical temperature on the NbN film thickness is shown.

were rf cleaned for 10 min. The discharge current for the subsequent deposition of AlN was varied from 200 mA to 500 mA in steps of 100 mA. With a high partial nitrogen pressure, it was ensured that all AlN films were deposited well inside region III. To see the effect of the AlN buffer layer, 4.8 nm of NbN were deposited *ex-situ* in a well characterized NbN dc magnetron sputter system. Simultaneously, an untreated, high-resistive Si[111] sample was included in the deposition, serving as a reference to the Si-AlN substrates. In Fig. 4.14 (a) the obtained critical temperatures of the NbN films are plotted over the discharge current of the AlN films. Here, the sample at $I_{dis} = 0$ mA corresponds to reference sample on pure Si[111]. As can be seen, all NbN films on an AlN buffer layer exhibit a significantly higher T_c than the one deposited directly on silicon. A maximum can be observed for $I_{dis,AlN} = 400$ mA with an increase of the critical temperature of $\delta T_c = 1.8$ K over the reference sample. Additionally, the resistivity of all samples deposited on AlN is lower as compared to the reference. The corresponding $R(T)$ measurements are shown in the inset of Fig. 4.14 (a).

Following the *ex-situ* deposition, *in-situ* bilayers were fabricated in the new deposition system. In Fig. 4.14 (b) the results, depicted as red solid symbols are compared to NbN films deposited directly on high resistive (high-ρ_0) Si[111] (black symbols). From the guide to the eyes, dashed lines in Fig. 4.14 (b), it can be seen that the increase in the critical temperature with respect to the NbN films deposited on pure silicon $\delta T_c \approx 3$ K is constant over all depicted NbN thicknesses. The higher film quality of NbN on Si-AlN bilayers (compared to Si) is most likely due to an effective increase in the film thickness, since the initial atomic layers remain undamaged due to the protective AlN layer. The AlN seems to result in a diffusion stop layer, preventing the degradation of the NbN at the interface between the film and the silicon substrate. Nevertheless, the significant δT_c is an encouraging result, allowing

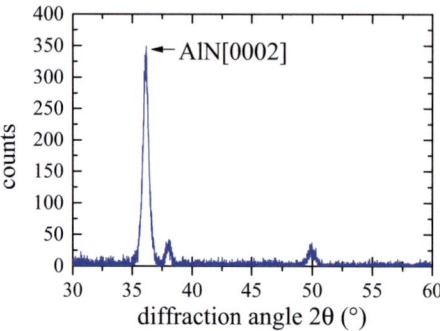

Figure 4.15.: X-ray diffraction measurement of a 148 nm thick AlN film on Si[111] substrate. The measurement has been performed in the grazing incident mode.

the use of less expensive silicon substrate, rather than $Al_2O_3[1\bar{1}02]$ and taking advantage of the high transparency of high-ρ_0 silicon in the THz and infrared frequency regimes.

In order to check the usability as a buffer layer for high frequency NbN based detector devices, the THz transparency was investigated. For this, the radiation absorption at a frequency of $f_{sig} = 650\,GHz$ was compared to that of an untreated high-ρ_0 Si[111] and a high-ρ_0 Si[111]-AlN substrate. The fact that no decrease in the transmission could be observed for the Si-AlN substrate, indicates that there are virtually no free charge carriers in the AlN layer, which could potentially absorb much of the high frequency radiation.

Finally, a 148 nm thick AlN film on Si[111] was examined by X-ray diffraction (XRD) measurements at the *Institute of Materials for Electrical and Electronic Engineering* (IWE) at the KIT. The measurements were performed by grazing incident detection (GID) under an incident angle of 1.2°. Thorough adjustment of the beamline allowed exclusive extraction of the AlN diffraction angle, without excitation of the silicon substrate. The corresponding spectrum is depicted in Fig. 4.15, showing a clear diffraction peak at $2\Theta \approx 36°$. The spectrum clearly indicates a dominant wurtzite hexagonal growth of the AlN layer and coincides very well with experiments found in literature [205, 207, 208]. In this particular orientation a lattice match to NbN[111] is promoted.

The AlN buffer layer has the additional advantage of being an absolute etch stop layer for reactive-ion etching using a CF_4-O_2 gas mixture. Even layers of thicknesses down to approximately 1 nm were not etched through after 1 min of etching time. However, particular care needs to be taken when the fabrication process includes development in KOH solutions. Similarly to results found in literature [213], the etching rate found here is also very high, reaching up to $1\,nm/s$ in developer. The advantages of AlN for the fabrication of NbN devices based on silicon substrates remain undeniable. In particular the significant increase of the critical temperature of up to $\delta T_c \approx 3\,K$ is unparalleled. The achieved T_cs for ultra-thin films of NbN on AlN exceed the values found in literature [166, 198], and are ideal for use in HEB mixer elements [Mar13].

Conclusion

From the investigations described above, deposition parameters of aluminum nitride films suitable for the incorporation in NbN devices on silicon substrates have been found. Thorough characterization of the discharge characteristics showed that tuning of the resulting AlN quality can be achieved in a reproducible manner. In particular the $\Delta p_{N2}(p_{N2})$ characteristics have proven to be a stable indicator for different regimes on the films. Introduction of a sufficiently high partial nitrogen pressure, allows fabrication of optically transparent AlN films exhibiting refractive indices at the order of $n \approx 2$, coinciding well with values found in literature [212]. The films deposited in region III of $\Delta p_{N2}(p_{N2})$ dependence furthermore are highly resistive, making it a promising candidate for a buffer layer underneath an ultra-thin film of NbN for hot-electron bolometers [Mar13]. The high resistivity was further validated by measurement of the THz transmission, where no change could be observed between the transmission through a high-ρ_0-Si substrate and a substrate with an additional AlN buffer layer. XRD measurements exhibited a clear diffraction peak at $2\Theta \approx 36°$, indicating a dominant wurtzite hexagonal growth of the AlN, which should also promote a hexagonal structure of NbN films deposited on top of this AlN buffer layer.

4.4. All-Niobium Nitride Josephson Junctions

As discussed above, conventional fabrication techniques make increasing film thicknesses with increasing number of underlying layers necessary. Thus lateral dimensions in layers patterned using lift-off are strongly limited. In the case of NbN/AlN/NbN multilayers, fabrication of Josephson junctions is even more challenging than for Nb/Al-AlO$_x$/Nb due to the rather thick electrodes at the order of the London penetration depth. Based on the conventional process, integration of layers subsequent to the patterning of the bottom electrode cannot be done using lift-off techniques unless very thick resists are used. Consequently, for NbN/AlN/NbN JJs planarized processes are required. The fully refined process discussed in section 3.2 is suitable for patterning of thick resists, without restrictions to lateral dimensions. However, the drawback of this process is that it is rather time-consuming, having a turn-around time of approximately eight days.

For initial characterization of newly developed trilayer deposition procedures a faster process is advantageous. Although fabrication of JJs using a cross-type fabrication process is very fast, the process is strongly limited in the lateral design and is not planar. Instead, a new process has been developed, which does not suffer from such limitations. It is based on merely four lithography steps and has a turn-around time of only three days. Fig. 4.16 shows the process flow for the edge type process. (a) shows the first lithography step, opening long trenches across the entire chip. Once etched, a small region of the resist is removed for contacting the Nb counter electrode for anodic oxidation. As a result of this the sidewalls of the trenches transformed into insulation Nb$_2$O$_5$, before SiO is thermally evaporated onto

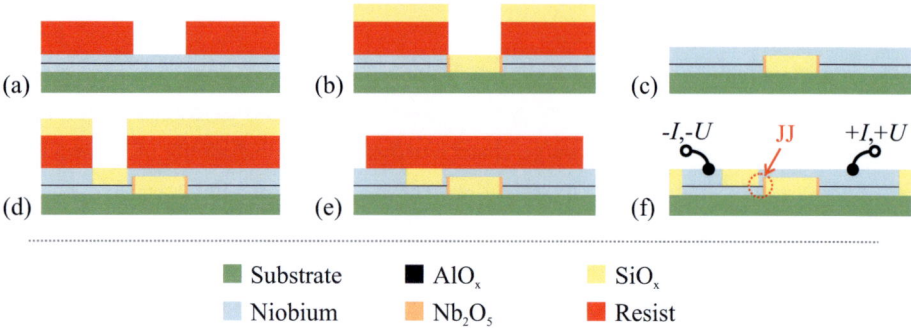

Figure 4.16.: Schematic cross-sectional representation of the edge type process. (a) Depicts the initial etching of the trenches with the self-aligned deposited SiO. (d) Shows the junction definition and (f) the separation of the individual structures on chip. The complete process is discussed in the text.

the chip, leveling the surface, *c.f.* (b). After stripping of the resist, the second lithography (not shown here), keeps the alignment marks in the corners of the chip covered, before a $100 - 200\,$nm thick Nb or NbN layer is dc sputtered, representing the wiring layer, as shown in (c). In the third lithography step, shown in (d), the junction area is defined by the overlap over the previously etched trench. At this point the wiring and the top electrode of the trilayer are etched in turn using RIE. If required for later processing, the new trench may be leveled using a SiO planarization layer. The final lithography, separating all structures from each other is shown in (e). Just like for the etching of the trilayer, again a sequence of RIE - IBE - RIE is employed, and the final topography may also be leveled using SiO. In (f) the final cross-section is depicted, with the electrical *IV*-connections indicated in black and the JJ itself highlighted by a dashed red circle. It should be noted that placing the electrical connections by ultra-sonic wiring bonding destroys any tunneling barrier directly underneath the point of connection. Along with the significantly larger area of the bond pad, as compared to the JJ, this ensures that only the highlighted SIS structure is measured instead of a serial connection of 2 JJs.

In combination with a mask consisting of 150 JJ structures, including single JJs of different sizes, long JJs and SQUIDS, this process allows fast characterization of trilayers, and is thus ideally suited for investigation of new trilayer processes. Additionally, it is suitable for patterning of thick trilayers, since it exhibits planar structures at intermediate process steps and may even be extended by a final self-aligned deposition of SiO, resulting in planar chip topography.

Using this process, NbN/AlN/NbN trilayers were characterized on different substrates. Concerning the substrate choice, it is necessary to take the selectivity during etching of the bottom NbN electrode to the substrate material into account. As has been mentioned above, pure silicon is etched strongly in a CF_4-O_2-plasma, making an additional etch stop layer

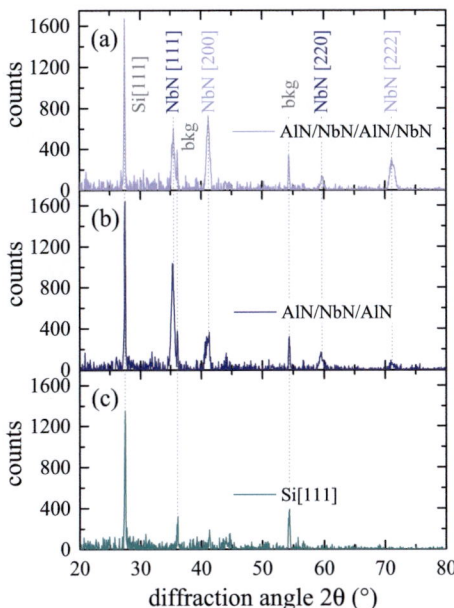

Figure 4.17.: XRD measurements of NbN/AlN/NbN multilayer (a) and the corresponding measurements of only the lower electrode on Si[111]/AlN (b) and the reference measurement of pure Si[111] substrate in (c).

necessary, whereas the selectivity for Si-SiO$_2$, Al$_2$O$_3$[1$\bar{1}$02] and MgO substrates is sufficient for a direct deposition. In the following, JJs fabricated on silicon substrate therefore refers to a process based on a Si-substrate with an additional AlN layer. The complete deposition sequence for a NbN/AlN/NbN trilayer on Si substrate consequently consists of an initial 10 min rf pre-cleaning, optional heating to T_{dep}, deposition of a thin AlN layer, deposition of the NbN bottom electrode, formation of the AlN tunneling barrier and a final deposition of the NbN counter electrode. In the case of Si-SiO$_2$, Al$_2$O$_3$[1$\bar{1}$02] and MgO substrates the process is done without the pre-cleaning and the first AlN layer.

Initially, the AlN etch stop layer thickness was varied in the range of $d_{AlN} \approx 5-60$ nm, investigating the influence on the subsequently deposited NbN. $d_{AlN} = 5$ nm was chosen as the minimum value to avoid any effects of dislocations that might be found within the first few atomic layers of AlN at the Si[111]-AlN interface, as has been reported widely [205–208]. Since no influence of the thickness on the NbN quality was found, d_{AlN} was thus fixed to roughly $10-15$ nm in most cases, keeping the total deposition time small. Fig. 4.17 shows three different X-ray diffraction (XRD) measurements that were done at the *Institute of Materials for Electrical and Electronic Engineering* at the KIT in Karlsruhe, Germany. The graph in (c) shows a scan of an untreated Si[111] substrate, serving as a reference to subsequent measurements of the bottom electrode including the AlN buffer

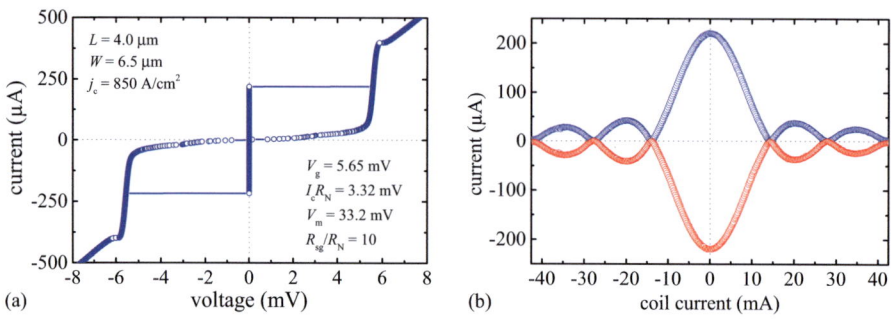

Figure 4.18.: In (a) the *IVC* and in (b) the $I_c(H)$ measurement of a NbN/AlN/NbN Josephson junction on MgO substrate is shown. The good quality parameters listed in (a) indicate the suitability of the trilayer deposition process for fabrication of JJs.

layer and tunneling barrier (b) and to a measurement of the final NbN/AlN/NbN trilayer (a). The first scan shows a clear diffraction peak at $2\Theta \approx 28°$, corresponding to the single crystalline Si[111] substrate [205]. The additional peaks at $\sim 36°$ and $\sim 54°$ are due to the background of the measurement system and are visible in all XRD scans performed in this particular system. In the second graph the diffraction scan of the AlN/NbN/AlN multilayer for the bottom electrode is depicted, with the main peak visible at $2\Theta \approx 35°$, indicating a preferred [111]-orientation of the NbN. However, even in the bottom electrode, [200] at $2\Theta \approx 41°$ and [220] orientations $2\Theta \approx 59°$ can be seen, with the latter being insignificantly small as compared to the primary [111] orientation. The upper graph, showing the results for a complete NbN/AlN/NbN trilayer exhibits a strongly increased and broadened diffraction peak of the [200] orientation and an additional peak at $2\Theta \approx 71°$, indicating a [222] growth of the NbN top electrode. The different primary orientations result from the additional interfaces due to the AlN tunneling barrier for the top electrode. Despite the fact that these XRD measurements indicate polycrystalline NbN, a primary growth-direction can be inferred for both electrodes. Based on such promising results, JJs were fabricated on MgO, $Al_2O_3[1\bar{1}02]$ and Si[111] substrates.

For first tests of the suitability for JJs, trilayers were fabricated on MgO, taking advantage of the good lattice-match to NbN. The crystallinity of plain NbN films on MgO[100] was also checked by XRD measurements, which showed a clear diffraction peak for NbN[100]. In Fig. 4.18 (a) the *IVC* and the extracted quality parameters of such a JJ can be seen. The highest gap voltage achieved $V_g = 5.65\,\text{mV}$ is close to the theoretical maximum at $T = 4.2\,\text{K}$ and is roughly a factor of 2 higher than that of Nb/Al-AlO$_x$/Nb junctions. Combined with the maximum Ambegaokar-Baratoff parameter of $I_c R_N = 3.32\,\text{mV}$, $V_{m,max} = 42.36\,\text{mV}$ and a resistance maximum ratio of $R_{sg}/R_N = 14.38$ such junctions are promising candidates for high frequency applications. In (b) the $I_c(H)$ modulation is shown for positive and negative bias current. The lack of any asymmetries and the sharp minima indicate a high uniformity

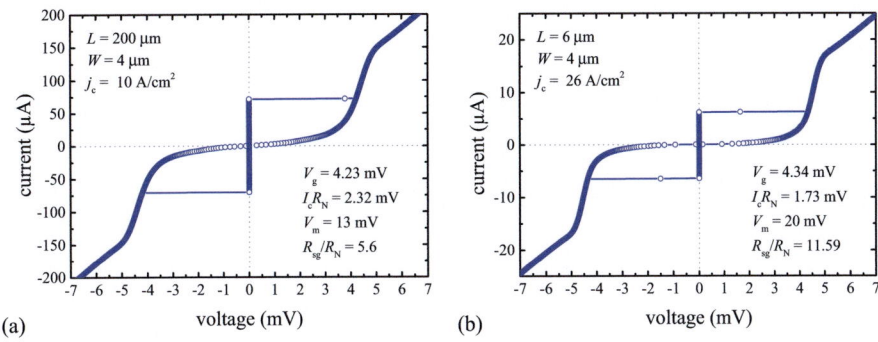

Figure 4.19.: The *IVC*s of NbN/AlN/NbN JJs on Al$_2$O$_3$[1$\bar{1}$02] and Si[111] substrate are shown in (a) and (b), respectively.

of the current distribution and thus also a good homogeneity of the tunneling barrier.

After having achieved a high junction quality for NbN/AlN/NbN trilayers on MgO[100], fabrication was also done on Al$_2$O$_3$[1$\bar{1}$02] and Si[111] substrates. Fig. 4.19 (a) shows an *IVC* of a geometrically long JJ on Al$_2$O$_3$[1$\bar{1}$02] and (b) of a short JJ on Si[111] with an AlN buffer layer. The quality parameters are again summarized in each graph. As expected, the gap voltage of both devices is lower than that on MgO. Surprisingly however, V_g is smaller on Al$_2$O$_3$[1$\bar{1}$02] than on Si[111]. This effect might be due to geometrical aspects of the first JJ. Due to the long lateral dimension $L = 200\,\mu$m, there is a higher probability of including spatial fluctuations of the gap voltage in the JJ area, as was found by Kirtley *et al.* [160]. Such variations in the gap could also explain the strong rounding of the *IVC* in the sub-gap region, which may be viewed as a parallel circuit of multiple gap voltages, resulting in a total reduction of V_g. Nevertheless, the quality of NbN/AlN/NbN JJs on Al$_2$O$_3$[1$\bar{1}$02] is also acceptable, exhibiting maximal gap voltages up to $V_g = 4.23\,$mV, $I_cR_N = 2.32\,$mV, $V_m = 13.17\,$mV and a resistance ratio of $R_{sg}/R_N = 7.3$.

Finally, Josephson junctions on Si[111] with an AlN buffer layer were investigated, see Fig. 4.19 (b). The junction quality is better than on Al$_2$O$_3$[1$\bar{1}$02] substrate, indicating that the observed difference in the surface topography discussed in section 4.2 does not influence the overall tunneling properties much. The maximum values of the Ambegaokar-Baratoff parameter was $I_cR_N = 2.13\,$mV, $V_m = 20.02\,$mV, $V_g = 4.38\,$mV and a resistance ratio of $R_{sg}/R_N = 11.59$. Furthermore, the gap voltage may be increased further by depositing electrodes of thicknesses in the range of $\lambda_L \approx 300\,$nm, which was extracted for thick NbN films on Si[111]. Higher critical-current densities than those presented in Fig. 4.18 and 4.19, necessary for mixer devices were also achieved and ranged up to $j_c = 3.7\,$kA/cm^2 on Si[111]. It should be noted that the gap voltages achieved so-far are sufficient for mixer elements in the THz-regime with $f_g > 1\,$THz and are higher than the value reported by Terai *et al.* [174, 175].

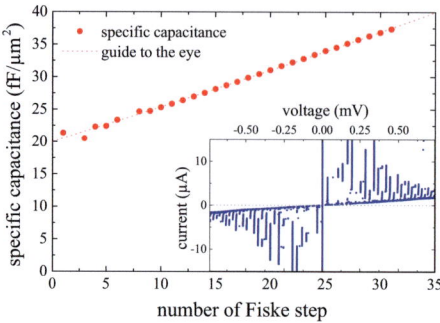

Figure 4.20.: Specific capacitance of a NbN/AlN/NbN trilayer on an $Al_2O_3[1\bar{1}02]$ substrate. The inset shows the corresponding measurement of the Fiske steps.

Approximating the London penetration depth to $\lambda_{L,NbN}^{GL} \approx 300\,nm$, the total magnetic thickness of the barrier results in $t_{ox,m} \approx 860\,nm$ according to Eq. 2.22, taking the physical barrier thickness to be $t_{ox} = 1.5\,nm$, top and bottom electrode to be $d_{t,b} = 210\,nm$ and the wiring layer to be $d_w = 150\,nm$. At a critical-current density of $j_c = 10\,A/cm^2$ of the $Al_2O_3[1\bar{1}02]$ sample, this results in a Josephson penetration depth of $\lambda_J = 55.3\,\mu m$. Under an externally applied magnetic field, Fiske steps were measured and the specific capacitance extracted according to Eq. 2.50. Fig. 4.20 shows the resulting dependence on the nth Fiske step. The values increase, due to the increasing losses. Compared to Nb/Al-AlO$_x$/Nb JJs the capacitance $C_{NbN/AlN/NbN}^{\star} \approx 20\,fF/\mu m^2$ is approximately a factor of 2 lower, which is comparable to values found in literature [214]. Such low capacities are highly beneficial for the design and consequently also the fabrication of SIS mixer elements and SQUIDs, since the intrinsic junction capacitance is significantly lower than that of devices with an AlO$_x$ tunneling barrier.

Conclusion

Based on the previous optimization of NbN films deposited directly on different substrates and the following optimization of AlN/NbN multilayers, trilayers of NbN/AlN/NbN were fabricated on MgO[100], $Al_2O_3[1\bar{1}02]$ and Si[111] substrates. The material composition as well as the primary orientation of the crystalline structure was investigated by means of X-ray diffraction. For film thicknesses of 150 nm for AlN a clear diffraction peak was found at $2\Theta = 36°$, corresponding to a [0001] growth direction. The crystallinity of the NbN layers for the JJ fabrication was measured for a multilayer of AlN/NbN/AlN on Si[111], without a top electrode deposited and a primary [111] orientation of the NbN bottom electrode was found. The top electrode exhibited more peaks indicating a more polycrystalline growth in the [220] and [200] orientation.

In terms of the gap voltages of the final NbN/AlN/NbN JJs, the obtained results do not yet reach the full potential of JJs based on NbN with an AlN tunneling barrier, they are however promising candidates for high frequency devices, since they all exhibit gap frequencies of $f_g > 1\,\mathrm{THz}$. The remaining quality parameters, like the Ambegaokar-Baratoff parameter $I_c R_N$, the characteristic voltage V_m and the resistance ratio R_{sg}/R_N are very good and in the case of Si[111] substrate even higher than found in literature [174, 175].

Finally, the specific capacitance was extracted for one of the trilayers. The low values of $C^\star_{\mathrm{NbN/AlN/NbN}} \approx 20\,\mathrm{fF}/\mu\mathrm{m}^2$ are very beneficial for design and fabrication of SIS mixer elements and SQUIDs, since the restrictions stemming from the inherent junction capacitance are somewhat relaxed as compared to those of Nb/Al-AlO$_x$/Nb trilayers. In summary it can be concluded, that a fully functional NbN/AlN/NbN fabrication process for a variety of substrates has been developed, exhibiting high quality parameters and critical-current densities of up to $j_c = 3.7\,\mathrm{kA/cm}^2$ on high-resistive Si[111] substrate. This technology is highly promising for future high frequency devices such as integrated THz-receivers or high speed RSFQ logic devices.

5. Summary

Within the framework of this thesis a large variety of Josephson junction devices has been designed, fabricated and investigated. Experiments on devices that have been developed using the conventional multilayer process of Nb/Al-AlO$_x$/Nb quickly outlined the limitations of the fabrication process, which has been available at the IMS prior to this thesis. Continuous refinement of the process, led to a new extremely versatile fabrication procedure of Josephson junctions. This resulted in a fully planar fabrication process up to the first wiring layer. Aside from the fact that a flat surface topography is achieved without the use of chemical mechanical polishing, the self-planarized process allows the integration of higher levels of superconducting metallization layers and also sub-μm patterning of all layers. Compared to the previously employed conventional process, the minimally achieved lateral junction dimensions have been reduced from $d_{JJ} \approx 700$ nm down to $d_{JJ} \approx 200$ nm using the newly developed self-planarized process [KMI$^+$11, MMB$^+$13]. In the highest metallization layer the minimum lateral dimension has been reduced from $3\,\mu$m to ~ 250 nm [Bue11, Bue13]. Furthermore, another fabrication process has been developed with respect to a minimum turnaround time, avoiding time consuming fabrication procedures and thus allowing for fast characterization of trilayers. This process consists of a maximum of four lithography steps and has a turnaround time of only three days.

Based on Nb/Al-AlO$_x$/Nb Josephson junctions, different types of Superconducting Quantum Interference Devices have been developed. The 2-JJ dc SQUID has been investigated in particular, focusing on the effects of asymmetries in the shunt resistance. By removing one of the two typically used shunts and replacing the other by a shunt of only half the resistance $R^{-1} = R_1^{-1} + R_2^{-1}$, the $V(\Phi_a)$ characteristics become asymmetric with one shallow slope and one very steep slope exhibiting up to $V_\Phi = 1.2\,\text{mV}/\Phi_0$. From a direct comparison of such an asymmetric SQUID with its symmetric counterpart, it was possible to show that asymmetries in the shunt resistance may be used to enhance the normalized energy resolution by a factor of 3.4 down to $e_r = 0.52$, representing the lowest value measured to this day [139, 140][RNM$^+$12]. Additionally, current asymmetric SQUIDs based on a design consisting of three junctions have been developed for the read-out of fractional vortex devices.

Based on preliminary studies together with the Physikalisches Institut Experimentalphysik II at Universität Tübingen, Germany, new long Josephson junction devices have been developed for the investigation of fractional vortices ($\wp$-vortices) and the dynamics thereof. In addition to fundamental investigations of $\wp$-vortices, including vortex activation mechanisms and spectroscopic measurements on single vortices, spectroscopy on vortex molecules has also been performed [85]. This has made it possible to experimentally show the increasing splitting of oscillatory molecule eigenmodes with decreasing distance between the fractional vortices [KMB+12]. The challenging design requirements initiated the above mentioned refinement of the JJ fabrication process, since the goal was to investigate $\wp$-devices not only in the classical regime at a bath temperature of $T = 4.2\,\text{K}$ but also in the quantum regime below the crossover temperature in the mK regime. High current density devices of sub-µm width and nano injectors have successfully been tested at temperatures down to $T = 25\,\text{mK}$. A dependence of the crossover temperature on the size of the artificial $\wp$-vortex has been observed, however the results are not yet conclusive. Further design iterations and experiments at mK temperatures are ongoing.

Together with the Institute of Radioengineering and Electronics (IRE) of the Russian Academy of Sciences (RAS), Moscow, Russia and the Physics Institute at the Technical University of Denmark (DTU), Lyngby, Denmark, improved on-chip oscillators based on long Josephson junctions have been investigated. These so-called flux-flow oscillators have been fabricated along with Josephson mixers, embedded in an rf matching circuit, using an aluminum hard mask process [KMI+11]. Furthermore, it was possible to show that the linewidth of the emitted electro-magnetic radiation from the FFO, can be reduced, by correlating the fluctuation afflicted bias current and control-line current for the creation of the magnetic field. This correlation has been realized by an LC-shunt, designed to have a high-pass characteristic, which does not interfere with the phase-locked loop mode of the FFO biasing. At a FFO frequency of $f_{\text{FFO}} = 237 - 238\,\text{GHz}$, a reduction of the linewidth by a factor of 1.8 has been shown. Averaged over more than 20 measurements in the frequency range from 150 GHz to 350 GHz a reduction of the linewidth of roughly 0.2 MHz has been observed.

The limitations of the existing Nb/Al-AlO$_x$/Nb technology initiated the development of a new 3-chamber vacuum sputter system, particularly designed for NbN/AlN/NbN Josephson junctions. The system consists of a load-lock chamber with integrated rf pre-cleaning and a transfer stage to the deposition chambers. The first deposition chamber is equipped with a 3″ magnetron and is used for reactive dc sputtering of NbN in an argon and nitrogen gas mixture. The substrate table is heatable up to 850 °C, while being movable at the same time. This ensures a stable temperature when transferring the sample to the AlN chamber and back. The second deposition chamber is also equipped with a 3″ magnetron and is used for dc sputtering of AlN, again in an atmosphere of argon and nitrogen.

Using the new multi-chamber system, NbN films have been optimized on various substrates including MgO[100], Al$_2$O$_3$[1$\bar{1}$02] and Si[111]. In the case of high-resistive Si[111] substrates an AlN buffer layer underneath the NbN layer has been employed to enhance the film qualities of the superconducting NbN. This has made it possible to increase the critical temperature of thin NbN films on silicon by approximately 3 K as compared to film deposited directly on the silicon surface. The obtained T_cs are the highest values reported on silicon. Such AlN/NbN bilayers are ideal candidates for the development of high-performance hot-electron bolometers (HEBs) and superconducting single-photon detectors (SNSPDs), due to their excellent quality parameters and the high transparency of the high-ρ_0 silicon in the THZ and infrared range [Mar13].

Thorough investigation of the surface morphology and topography of NbN, deposited under various different conditions, allowed fabrication of films with smooth surfaces and high critical temperatures. The surface has been investigated by means of scanning electron microscopy (SEM) as well as atomic force microscopy. It has been found that for power densities of $P \gtrsim 4\,\mathrm{W/cm^2}$, the deposited NbN films exhibit root mean square surface roughnesses of less than 1 unit cell on both, $Al_2O_3[1\bar{1}02]$ and Si[111] substrates. Additionally, the percental nitrogen consumption during deposition has been found to be a global indicator for the superconducting film quality, for a specific deposition system. Based on these findings, NbN/AlN/NbN multilayers have been deposited for the fabrication of Josephson junctions. The crystallinity of the individual NbN layers have been checked by means of X-ray diffraction measurements, and a primary [111]-orientation for the bottom electrode and [200] for the top electrode has been found. The AlN crystallinity has been investigated using a specifically deposited thick layer and a clear diffraction peak at $2\Theta = 36°$ has been found, indicating an almost monocrystalline wurtzite hexagonal growth.

The results obtained on NbN/AlN/NbN JJs are very promising for high frequency devices in the THz range, since the achieved gap voltages on each substrate exceeded $V_g = 4.13\,\mathrm{mV}$, which corresponds to the minimum gap voltage required for mixing of frequencies $f \geq 1\,\mathrm{THz}$. Based on MgO substrate, the maximum gap voltage achieved so far at the IMS has been $V_g = 5.65\,\mathrm{mV}$, which is comparable to the highest values reported in literature [101, 159, 162, 170]. However, the junction quality on Si[111] is particularly remarkable, being higher than of those devices reported in literature up to this date [174, 175]. Based on this NbN/AlN/NbN process, with JJs having critical-current densities in the range of a few $\mathrm{kA/cm^2}$, the development of highly sophisticated devices is possible, due to an almost fully planar process, allowing for sub-μm lateral dimensions in all layers. Further combining the newly developed LC-shunted FFO with possibly a HEB mixer device on a single high-ρ_0 Si[111] chip, may introduce a new generation of high-performance fully integrated heterodyne THz receiver.

A. Appendix

A.1. Photo Lithography

Positive Photo Lithography

At the IMS undiluted AZ5214E resist [215], spun at 7000 rpm for one minute, is typically used, resulting in a resist thickness of approximately 1400 nm. A 5 min soft-bake at 85 °C evaporates the thinner, leaving a photosensitive resist layer on top of the trilayer. The thick ridge, caused during spin coating is exposed first for 30 s and removed in MIF726 developer [215], resulting in an almost perfectly planar resist surface in the central $\sim 8.5 \times 8.5 \, \text{mm}^2$ part of the chip. The subsequent patterning in this region is performed using a chromium mask, a 10 s UV-exposure and development in MIF726.

Image Reversal Lithography

For the image reversal lithography, also the undiluted AZ5214E resist [215], spun at 7000 rpm for one minute, is typically used at the IMS. After removing of the thick ridge, the resist is exposed with the inverse image of the wanted pattern for 5 s, baked for 5 min at 120 °C and finally the entire chip is exposed for 1 min. This post-baking causes the broken polymer chain in the previously exposed areas to cross link, leaving these areas resistant to any further UV-exposure. For high contrast, a comparably weak developer MIF 300-47 [215] is used.

A.2. Etching and Deposition

Trilayer Deposition

A 2″ SiO_2 wafer is placed in the load-lock chamber and evacuated to a base pressure of approximately $3 \cdot 10^{-4}$ Pa. For cleaning of the SiO_2 surface, an Argon rf plasma is applied for 10 minutes prior to the deposition of the Nb bottom electrode.

The niobium deposition is power-controlled and carried out at a power of $P_{Nb} = 300 \, \text{W}$ in pure argon atmosphere at a pressure of $p_{Nb} = 0.96$ Pa. This results in a deposition rate of $dr_{Nb} = 0.75 \, \text{nm/s}$, forming a poly-crystalline niobium film as shown in Fig. 3.1. The following deposition of the aluminum wetting layer is power-controlled to $P_{Al} = 100 \, \text{W}$ at a pressure $p_{Al} = 0.72$ Pa, resulting in a deposition rate $dr_{Al} = 0.3 \, \text{nm/s}$. The top electrode, finalizing the trilayer, is again deposited with the identical parameters as the bottom electrode.

Anodic Oxidation

The anodic oxidation is performed in an aqueous solution of ammonium pentaborate $(NH_4)B_5O_6$ and ethylene glycol $C_2H_6O_2$ at a maximum applied current of $0.7\,mA$, for normally $5\,min$. The conversion rate of Nb to Nb_2O_5 is given by:

$$0.88\,nm \text{ of Nb} \xrightarrow{1\,\text{Volt}} 2.3\,nm \text{ of } Nb_2O_5. \tag{A.1}$$

The conversion rate of Al to AlO_x is given by:

$$0.9\,nm \text{ of Al} \xrightarrow{1\,\text{Volt}} 1.3\,nm \text{ of } AlO_x. \tag{A.2}$$

Deposition of Resistor Material

The sputter process for Palladium and Manganin is power-regulated to $P = 50\,W$ at an Ar-pressure of $p_{Pd} = 0.71\,Pa$, resulting in a deposition rate of $dr_{Pd} = 0.48\,nm/s$ for palladium and $dr_{Manganin} = 0.80\,nm/s$. The dependence of the square resistance $R_\square$ is depicted in Fig. 3.5.

Deposition of SiO

SiO is typically used for insulation of the JJs at the IMS. The SiO is thermally evaporated from granular material in a tantalum crucible. For heating a current of $I_{SiO} \approx 5.0 - 5.2\,A$ is applied, resulting in a deposition rate of $dr_{SiO} \approx 0.6 - 1.0\,nm/s$. The base and deposition pressure of the system is usually $p_{base} < 10^{-4}\,Pa$, leading to a highly anisotropic deposition.

Deposition of the Niobium Wiring Layer

The niobium wiring layer is deposited using a $2''$ dc magnetron after a $10\,min$ pre-cleaning procedure, employing an argon ion gun. The pre-cleaning is typically performed at a current of $I_{pc} = 10\,mA$, a voltage of $V_{pc} = 100\,V$ and at a pressure of $p_{pc} = 0.13\,Pa$. By this approximately $3 - 4\,nm$ of Nb is etched, ensuring low contact resistivity.

The niobium deposition is a current-regulated dc sputter process at $I_{dis,Nb} = 175\,mA$ in pure argon atmosphere $p_{Ar} = 0.5\,Pa$, resulting in a discharge voltage of $V_{Nb} \approx 280\,V$. For deposition of the $400\,nm$ thick Nb wiring layer, the deposition is typically split in two turns of a maximum $13\,min$ deposition time, in order to prevent over-heating of the resist. This corresponds to a deposition rate of $dr_{Nb} \approx 0.25\,nm/s$.

Reactive-Ion Etching

At the IMS, reactive-ion etching is typically performed using a gas mixture of CF_4 and O_2, at respective flows of 30 sccm and 5.9 sccm. At a process pressure of $p_{RIE} = 34.7$ Pa and an applied rf power of $P_{RIE} = 100$ W, this leads to an etching rate of niobium of $er_{Nb} \approx 3$ nm/s. Comparable values have also been found for the etching of NbN. The exact etching rate of NbN depends however on the stoichiometry of the particular NbN films.

Ion-Beam Etching

Ion beam etching at the IMS is manually current-regulated to $I_{IBE} = 48 \pm 1$ mA. At an argon flow of $\mathscr{F}_{Ar} = 6$ sccm, resulting in a process pressure of $p_{Ar} = 1.5 \cdot 10^{-2}$ Pa, a positive grid-voltage of $V_+ = 250$ V, a negative grid-voltage of $V_- = -200$ V and an approximate rf power of $P_{IBE} \approx 130$ W, this results in an etching rate of $er_{AlOx} \gtrsim 0.008$ nm/s for AlO_x and $er_{Nb} \gtrsim 0.05$ nm/s for Nb.

List of Figures

List of Tables

Nomenclature

α	Dimensionless damping parameter $(=\frac{1}{\sqrt{\beta_c}})$
α_C	Capacitance-asymmetry parameter in a SQUID
α_I	Current-asymmetry parameter in a SQUID
α_L	Inductance-asymmetry parameter in a SQUID
α_R	Resistance-asymmetry parameter in a SQUID
α_t	Pre-factor in the thermal escape rate
$\bar{x}$	Mean switching current of a JJ-array (A)
β_c	Stewart-McCumber parameter $(=\frac{2\pi I_c R^2 C}{\Phi_0})$
β_L	Screening parameter of a SQUID $(=\frac{2I_0 L_{sq}}{\Phi_0})$
Δf	Emission linewidth of a flux-flow oscillator (Hz)
ΔI_{SIS}	Current increase in the harmonic mixer at V_p under the influence of f_p (A)
Δp_{N2}	Nitrogen consumption (Pa)
Δs	Distance between two narrow bias lines (m)
δT_c	Width of the superconducting transition (K)
Δ	Energy gap of a superconductor (eV)
$\Delta\mathcal{U}$	Potential barrier between two qubit states (eV)
ε	Energy resolution of a SQUID $(=\frac{S_\Phi}{2L_{sq}})$
ε_{ins}	Relative permittivity of the insulating layer
η	Damping constant in the tilted washboard analogy $(=\frac{\Phi_0}{2\pi}\frac{1}{R})$
Γ	Escape rate from a potential well (Hz)
γ	Normalized current of a Josephson junction $(=\frac{I}{I_c})$
$\gamma_c(\wp)$	Normalized depinning current of a fractional vortex
Γ_N	Noise ratio in a SQUID $(=\frac{I_N}{I_0})$
γ_N	Normalized noise current of a Josephson junction $(=\frac{I_N}{I_c})$
Γ_p	Damping constant in the pendulum analogy $(=\frac{\Phi_0}{2\pi}\frac{1}{R})$
ι	Current distribution in a Josephson junction $(\frac{A}{m})$
ι_0	Mean linear current density $(\frac{A}{m})$
κ	Phase discontinuity caused by a fractional vortex
κ_c	Critical value of κ at which the vortex changes from its direct state to its complementary state
κ_{GL}	Ginzburg-Landau parameter
λ	Wavelength (m)
$\lambda_{J,eff}$	Effective Josephson penetration depth (m)

λ_J	Josephson penetration depth (m)
$\lambda_{L,d}^{GL}$	Ginzburg-Landau estimation of the London penetration depth in the dirty limit (m)
λ_L	London penetration depth (m)
$\langle V \rangle$	Mean voltage over time ($= \dot{\varphi}\frac{\Phi_0}{2\pi}$) (V)
$\mathbb{Z}$	Set of integers
$\mathscr{F}_x$	Flow of gas x (sccm)
$\mathscr{L}$	Mean circumference of an annular Josephson junction ($\mathscr{L} = 2\pi r$) (m)
$\mathscr{A}$	Amplitude (a.u.)
$\mu(\tilde{z},\tilde{t})$	Continuous magnetic term of the Josephson phase
μ_0	Vacuum permeability ($4\pi \cdot 10^{-7}\frac{H}{m}$)
$\omega(k,\gamma)$	Dispersion relation
$\omega_0(\wp,\gamma)$	Oscillatory eigenfrequency of a fractional vortex (Hz)
ω_{ext}	Externally applied microwave radiation (Hz)
ω_{pl}	Plasma frequency of a Josephson junction (Hz)
ω_+	In-phase oscillatory eigenmode of a fractional-vortex molecule (Hz)
ω_-	Out-of-phase oscillatory eigenmode of a fractional-vortex molecule (Hz)
Φ	Magnetic flux (Wb)
Φ_0	Magnetic flux quantum ($= \frac{h}{2e} = 2.07 \cdot 10^{-15}$ Wb)
Φ_a	Magnetic flux inside a superconducting ring structure ($= n \cdot \Phi_0$) (Wb)
Φ_{cl}	Magnetic flux caused by a control-line current (Wb)
Φ_{ext}	Externally applied magnetic flux (Wb)
π-JJ	Josephson junction with a π phase shift
Ψ	Superconducting wavefunction
ρ	Resistivity ($\Omega \cdot$m)
σ	Standard deviation of a switching histogram (A)
τ_c	Characteristic time of a Josephson junction ($= \frac{\Phi_0}{2\pi I_c R_{(N)}}$)
$\Theta(\tilde{z})$	Step function ($= \kappa H(\tilde{z})$)
Θ_p	Torque of inertia in the pendulum analogy ($= \frac{\Phi_0 C}{2\pi}$)
$\tilde{t}$	Normalized time ($= \omega_{pl} \cdot t$)
$\tilde{z}$	Normalized spatial coordinate ($= \frac{z}{\lambda_J}$)
φ	Gauge-invariant phase difference across a Josephson junction
φ_p	Angle of twist of a pendulum ($^\circ$)
φ_{sc}	Phase of the superconducting wavefunction
φ_w	Coordinate in the tilted washboard analogy (m)
$\vec{\nabla}$	Nabla operator
$\vec{A}$	Magnetic vector potential ($\frac{V \cdot s}{m}$)
$\vec{B}_{cl}$	Magnetic flux density due to control-line current (T)
$\vec{B}$	Magnetic flux density (T)

$\vec{E}$	Electric field ($\frac{V}{m}$)
$\vec{F}_L$	Lorentz-force ($\frac{kg \cdot m}{s^2}$)
$\vec{k}$	Wave vector
$\wp$	Topological charge of a fractional vortex
$\xi(T)$	Ginzburg-Landau coherence length (m)
ξ_0	BCS coherence length (m)
ζ	Sheet resistance ($= R \cdot W \Delta z$)
ζ_s	Surface sheet resistance ($= R_s \cdot W \Delta z$)
A	Area of a Josephson junction (m^2)
a	Distance between two phase discontinuity points (m)
a_c	Critical distance between two phase discontinuities (m)
A_t	Target area (m^2)
a_x	Lattice constant a of a given material x (m)
b	Exponent in fit of critical-current density
B_a	Externally applied magnetic field (T)
B_{rem}	Remanent magnetic field (T)
C	Capacitance (F)
C_{shunt}	Shunt capacitance (F)
C_{sq}	SQUID capacitance ($= \frac{C_1 + C_2}{2}$) (F)
c_{sw}	Swihart velocity ($\frac{m}{s}$)
c_x	Lattice constant c of a given material x (m)
$C^\star$	Specific capacitance ($\frac{F}{m^2}$)
d	Thickness (m)
d_{JJ}	Diameter of a Josephson junction (m)
d_{via}	Diameter of a vertical interconnect access (m)
dr	Deposition rate ($\frac{m}{s}$)
dZ	Distance between two dc current injectors (m)
E	Potential energy (eV)
e	Elementary charge ($= 1.662 \cdot 10^{-19}$ C)
E_{ext}	Externally applied energy, $e.g.$ in form of microwave radiation (eV)
E_g	Direct band gap of a semiconducting / insulating material (eV)
E_J	Josephson coupling energy ($= \frac{\Phi_0 I_c}{2\pi}$)
E_{ox}	Oxygen Exposure (Pa $\cdot$ s)
e_r	Normalized energy resolution of a SQUID
E_{th}	Thermal energy ($= k_B T$)
er	Etching rate ($\frac{m}{s}$)
f	Frequency (Hz)
f_{FFO}	Emission frequency of a flux-flow oscillator (Hz)
f_g	Gap frequency (Hz)

f_{IF}	Intermediate frequency (Hz)
f_J	Josephson frequency ($= 483.5979 \frac{GHz}{mV}$)
f_{lim}	Limiting frequency of a given superconducting multilayer technology (Hz)
f_{LO}	Local oscillator frequency (Hz)
f_{mix}	Mixed frequency (Hz)
f_p	Pumping frequency (Hz)
f_{sig}	Signal frequency (Hz)
G	Gain of an amplifier
g	Gravitation constant ($= 6.674 \cdot 10^{-11} \frac{m^3}{kg \cdot s^2}$)
h	Planck's constant ($= \hbar \cdot 2\pi = 6.62 \cdot 10^{-34}\,J \cdot s$)
$H(\tilde{z})$	Heaviside function
H_c	Critical field of a superconductor ($\frac{A}{m}$)
I	Current (A)
i	Normalized bias current of a SQUID ($= \frac{I}{I_0}$)
I_0	Half of the critical current of a SQUID ($= \frac{I_{c,1}+I_{c,2}}{2}$) (A)
I_b	Bias current (A)
$I_{c,0}$	Critical current of a Josephson junction at $T = 0\,K$ (A)
$I_{c,sw}$	Switching current of a Josephson junction (A)
I_{cap}	Displacement current in the RCSJ model (A)
I_{circ}	Circulating current in a superconducting ring structure for $\Phi_a \neq n \cdot \Phi_0$ (A)
i_{circ}	Normalized circulating current in a SQUID ($= \frac{I_{circ}}{I_0}$)
I_{cl}	Control-line current (A)
I_{coil}	Coil current (A)
I_{cs}	Current from current source in SQUID measurement (A)
I_c	Critical current of a Josephson junction (A)
I_{dis}	Discharge current (A)
I_{FFO}	Bias current of a flux-flow oscillator (A)
I_f	Feedback current (A)
I_{HM}	Bias current of a harmonic mixer (A)
I_{inj}	Injector current (A)
I_m	Mean switching current of a switching histogram of a Josephson junction (A)
$i_{n,theory}$	Theoretical normalized critical current for a homogeneous bias current distribution in a long Josephson junction ($\frac{I_{c,theory}}{j_c W \lambda_J}$)
I_N	Noise current in the RCSJ model (A)
i_n	Normalized critical current of a long Josephson junction ($\frac{I_c}{j_c W \lambda_J}$)
I_R	Quasiparticle current in the RCSJ model (A)
I_r	Re-trapping current (A)
$I_{s,max}$	Maximum supercurrent carried by a superconducting system (A)
I_s	Supercurrent in a Josephson junction (A)

IVC	Current voltage characteristic
$j_{c,mean}$	Mean critical-current density extracted from the IVC of a JJ-array ($\frac{A}{cm^2}$)
j_c	Critical-current density of a trilayer ($\frac{A}{cm^2}$)
$j_{s,c}$	Critical-current density of a superconductor ($\frac{A}{cm^2}$)
j_s	Supercurrent density in a superconductor ($\frac{A}{cm^2}$)
k_B	Boltzmann constant ($= 1.38 \cdot 10^{-23} \frac{kg \cdot m^2}{s^2 \cdot K}$)
L	Length of a Josephson junction (m)
l	Normalized length of a Josephson junction ($= \frac{L}{\lambda_J}$)
L_i	Input inductance (H)
L_{J0}	Josephson inductance for $I \ll I_c$ ($= \frac{\Phi_0}{2\pi I_c}$)
L_J	Inductance of a Josephson junction (H)
L_p	Parasitic inductance (H)
l_p	Length of a pendulum (m)
L_{sq}	SQUID inductance ($= L_1 + L_2$) (H)
L_s	Inductance in the extended RCSJ model for long Josephson junctions (H)
L_W	Length of an inductance in the wiring layer (m)
$L^\star$	Specific inductance ($= \frac{\mu_0 d}{W}$)
m	Integer
$m_{A\text{-}B}$	Mismatch between lattice of material A and B (%)
M_f	Mutual inductance of the feedback coil to the SQUID amplifier (H)
M_i	Mutual inductance of the input coil to the SQUID amplifier (H)
M_p	External driving torque of a pendulum ($\frac{kg \cdot m^2}{s^2}$)
m_p	Mass of a pendulum (kg)
m_s	Mass of a superconducting charge carrier (kg)
m_w	Mass of the particle in the tilted washboard analogy ($= \frac{\Phi_0}{2\pi} C$)
n	Integer
n_s	Cooper pair density ($\frac{1}{cm^3}$)
P	Power density ($\frac{W}{m^2}$)
p	Pressure (Pa)
p_{Ar}	Partial argon pressure (Pa)
p_{dis}	Pressure during plasma discharge (Pa)
p_{N2}	Partial nitrogen pressure (Pa)
p_{oxy}	Oxidation pressure (Pa)
P_{rf}	Applied rf power (W)
Q	Quality factor of a Josephson junction ($= \omega_{pl} RC$)
q	Electric charge (C)
R	Resistance (Ω)
r	Radius (m)
R_d	Differential resistance of a Josephson junction ($= \frac{\partial V_b}{\partial I_b}$) ($\Omega$)

R_d^{cl}	Differential transresistance of a Josephson junction with respect to the control-line current ($= \frac{\partial V_b}{\partial I_{cl}}$) ($\Omega$)
R_f	Resistance in feedback circuitry (Ω)
R_i	Serial resistance in the SQUID input coil (Ω)
R_N	Normal state resistance (Ω)
R_{sg}	Sub-gap resistance of a Josephson junction (Ω)
R_{shunt}	Shunt resistance (Ω)
R_{sq}	Twice the SQUID resistance ($= \frac{2R_1R_2}{R_1+R_2}$) ($\Omega$)
R_s	Surface resistance (Ω)
R_{tot}	Total resistance of a (shunted) Josephson junction (Ω)
$S_I(f)$	Nyquist noise ($\frac{V}{\sqrt{Hz}}$)
s_{JJ}	Linear shrinkage of Josephson junctions (m)
$S_V^{1/2}$	Voltage noise ($\frac{V}{\sqrt{Hz}}$)
S_Φ	Spectral density of equivalent flux noise ($\frac{Wb}{Hz}$)
T	Temperature (K)
t	Time (s)
T_{base}	Base temperature of a dilution fridge (K)
T_{bath}	Bath temperature (K)
T_c	Critical temperature of a superconductor (K)
T_{dep}	Deposition temperature (°C)
$t_{ins,eff}$	Effective thickness of the insulator in the idle region of a Josephson junction (m)
$t_{ox,eff}$	Effective barrier thickness in the active region of a Josephson junction (m)
$t_{ox,m}$	Magnetic barrier thickness of a Josephson junction (m)
t_{oxy}	Oxidation time (Pa)
T_{ox}	Oxidation temperature (°C)
t_{ox}	Physical thickness of the tunnel barrier (m)
T^*	Crossover temperature (K)
U	Potential barrier in the tilted washboard analogy (eV)
v	Normalized velocity of a moving fluxon in a long Josephson junction
V_{ao}	Applied voltage during anodic oxidation (V)
V_b	Bias voltage (V)
V_c	Ambegaokar-Baratoff parameter (V)
V_{FFO}	Voltage across a flux-flow oscillator (V)
V_{FF}	Voltage of flux-flow steps (V)
V_f	Feedback voltage (V)
V_g	Gap voltage ($= \frac{2\Delta}{e}$) (Ω)
V_{HM}	Voltage across a harmonic mixer (V)
V_m	Characteristic voltage of a Josephson junction ($= I_c R_{(N)}$)
V_n	Voltage of Fiske steps (V)

V_p	Bias voltage of the harmonic mixer (V)
V_{ZFS}	Voltage of zero field steps (V)
V_Φ	Transfer function of a SQUID ($\frac{V}{Wb}$)
W	Width of a Josephson junction (m)
w_1	Width of the active region of a Josephson junction (m)
w_2	Width of the idle region of an overlap Josephson junction (m)
W_{inj}	Width of a single dc current injector (m)
W_{pair}	Width of a dc current injector pair (m)
W_W	Width of an inductance in the wiring layer (m)
x	Spatial coordinate (m)
y	Spatial coordinate (m)
z	Spatial coordinate (m)
0-JJ	Conventional Josephson junction
A/D	Analog to digital
aAFM	Asymmetric anti-ferromagnetic state of a fractional-vortex molecule
ac	Alternating current
AFM	Anti-ferromagnetic state of a fractional-vortex molecule
aFM	Asymmetric ferromagnetic state of a fractional-vortex molecule
CMP	Chemical-mechanical polishing
D/A	Digital to analog
dc	Direct current
DUT	Device under Test
EBL	Electron-beam lithography
FFO	Flux-flow oscillator
FM	Ferromagnetic state of a fractional-vortex molecule
GEV	Generalized extreme value
GID	Grazing incident detection
HEB	Hot-electron bolometer
HM	Harmonic mixer
IBE	Ion-beam etching
IF	Intermediate frequency
JJ	Josephson junction
LHe	Liquid helium
LJJ	Long Josephson junction
LNA	Low noise amplifier
LO	Local oscillator
MQT	Macroscopic quantum tunneling
MW	Microwave
OL	Open loop mode of a SQUID amplifier

PCI	Peripheral component interconnect
PL	Photo lithography
PLL	Phase-locked loop
PMMA	Polymethyl methacrylate
PUL	Pick-up loop
RA	Resonant activation
RCSJ	Resistively and Capacitively Shunted junction model
rf	Radio frequency
RIE	Reactive-ion etching
RMS	Root mean square
RRR	Residual resistance ratio
RT	Room temperature
RTA	Room temperature amplifier
SA	SQUID amplifier
sAFM	Symmetric anti-ferromagnetic state of a fractional-vortex molecule
SC	Superconductor
SEM	Scanning Electron Microscope
sFM	Symmetric ferromagnetic state of a fractional-vortex molecule
SFS	Superconductor - Ferromagnet - Superconductor multilayer
SIS	Superconductor - Insulator - Superconductor multilayer
SMR	Superconducting microstrip resonator
SNSPD	Superconducting nanowire single-photon detector
SQUID	Superconducting quantum interference device
TA	Thermal activation
VCO	Voltage-controlled oscillator
via	Vertical interconnect access
VLL	Voltage-locked loop mode of a SQUID amplifier
XRD	X-ray diffraction

Bibliography

[1] B. D. Josephson. Possible new effects in superconductive tunnelling. *Physics Letters*, 1(7):251 – 253, 1962.

[2] S. Shapiro. Josephson Currents in Superconducting Tunneling: The Effect of Microwaves and Other Observations. *Phys. Rev. Lett.*, 11:80–82, Jul 1963.

[3] M. T. Levinsen, R. Y. Chiao, M. J. Feldman, and B. A. Tucker. An inverse ac Josephson effect voltage standard. *Applied Physics Letters*, 31(11):776–778, 1977.

[4] J. Niemeyer, J. H. Hinken, and R. L. Kautz. Microwave-induced constant-voltage steps at one volt from a series array of Josephson junctions. *Applied Physics Letters*, 45(4):478–480, 1984.

[5] C. A. Hamilton, R. L. Kautz, R. L. Steiner, and F. L. Lloyd. A practical Josephson voltage standard at 1 V. *Electron Device Letters, IEEE*, 6(12):623 – 625, Dec 1985.

[6] C. A. Hamilton, F. L. Lloyd, K. Chieh, and W. C. Goeke. A 10-V Josephson voltage standard. *Instrumentation and Measurement, IEEE Transactions on*, 38(2):314 –316, Apr 1989.

[7] K. K. Likharev. Rapid single-flux-quantum logic. In *The new superconducting Electronics*, pages 423–452. Springer, 1993.

[8] T. Nagatsuma, K. Enpuku, F. Irie, and K. Yoshida. Flux-flow type Josephson oscillator for millimeter and submillimeter wave region. *Journal of Applied Physics*, 54(6):3302–3309, 1983.

[9] T. Nagatsuma, K. Enpuku, K. Yoshida, and F. Irie. Flux-flow-type Josephson oscillator for millimeter and submillimeter wave region. II. Modeling. *Journal of Applied Physics*, 56(11):3284–3293, 1984.

[10] T. Nagatsuma and K. Enpuku and K. Sueoka and K. Yoshida and F. Irie. Flux-flow-type Josephson oscillator for millimeter and submillimeter wave region. III. oscillation stability. *Journal of Applied Physics*, 58(1):441–449, 1985.

[11] D. Winkler and T. Claeson. High-frequency limits of superconducting tunnel junction mixers. *Journal of Applied Physics*, 62(11):4482–4498, 1987.

[12] B. D. Jackson and T. M. Klapwijk. The current status of low-noise THz mixers based on SIS junctions. *Physica C: Superconductivity*, 372(0):368–373, 2002.

[13] C. D. Tesche and J. Clarke. dc SQUID: Noise and optimization. *Journal of Low Temperature Physics*, 29:301–331, 1977.

[14] J. Clarke and A. I. Braginski. *The SQUID Handbook: Fundamentals and Technology of SQUIDs and SQUID Systems*, volume 1. Wiley-VCH Verlag GmbH & Co. KGaA, 2004.

[15] K. K. Likharev and V. K. Semenov. RSFQ logic/memory family: A new Josephson-junction technology for sub-terahertz-clock-frequency digital systems. *Applied Superconductivity, IEEE Transactions on*, 1(1):3–28, 1991.

[16] E. Goldobin, D. Koelle, and R. Kleiner. Semifluxons in long Josephson $0 - \pi$-junctions. *Phys. Rev. B*, 66:100508, Sep 2002.

[17] A. V. Ustinov. Fluxon insertion into annular Josephson junctions. *Applied Physics Letters*, 80(17):3153–3155, 2002.

[18] D. van Delft and P. Kes. The discovery of superconductivity. *Physics Today*, 63(9):38–43, 2010.

[19] W. Buckel and R. Kleiner. *Supraleitung - Grundlagen und Anwendungen*, volume 6. WILEY-VCH Verlag GmbH & Co. KGaA Weinheim, 2004.

[20] F. London and H. London. The Electromagnetic Equations of the Supraconductor. *Proceedings of the Royal Society of London. Series A - Mathematical and Physical Sciences*, 149(866):71–88, 1935.

[21] J. Bardeen, L. N. Cooper, and J. R. Schrieffer. Theory of Superconductivity. *Phys. Rev.*, 108:1175–1204, Dec 1957.

[22] H. Fröhlich. Theory of the Superconducting State. I. The Ground State at the Absolute Zero of Temperature. *Phys. Rev.*, 79:845–856, Sep 1950.

[23] L. N. Cooper. Bound Electron Pairs in a Degenerate Fermi Gas. *Phys. Rev.*, 104:1189–1190, Nov 1956.

[24] W. S. Corak, B. B. Goodman, C. B. Satterthwaite, and A. Wexler. Exponential Temperature Dependence of the Electronic Specific Heat of Superconducting Vanadium. *Phys. Rev.*, 96:1442–1444, Dec 1954.

[25] W. S. Corak, B. B. Goodman, C. B. Satterthwaite, and A. Wexler. Atomic Heats of Normal and Superconducting Vanadium. *Phys. Rev.*, 102:656–661, May 1956.

[26] V. L. Ginzburg and L. D. Landau. . *Zh. Eksp. Teor. Fiz.*, 20:1064, 1950.

[27] M. Tinkham. *Introduction to Superconductivity*, volume Second Edition. Dover Publications Inc., Mineola, New York, 2004.

[28] A. A. Abrikosov. Translated: Magnetic properties of superconductors of the second group. *Translated: Sov. Phys. - JETP; Original: Zh. Eksp. Teor. Fiz.*, 5:1174, 1957.

[29] F. London. *Superfluids - Macroscopic Theory of Superconductivity*, volume 1. John Wiley & Sons, New York, Inc. - Chapman & Hall, London, Limited, 1950.

[30] B. S. Deaver and W. M. Fairbank. Experimental Evidence for Quantized Flux in Superconducting Cylinders. *Phys. Rev. Lett.*, 7:43–46, Jul 1961.

[31] R. Doll and M. Näbauer. Experimental Proof of Magnetic Flux Quantization in a Superconducting Ring. *Phys. Rev. Lett.*, 7:51–52, Jul 1961.

[32] P. W. Anderson and J. M. Rowell. Probable Observation of the Josephson Superconducting Tunneling Effect. *Phys. Rev. Lett.*, 10:230–232, Mar 1963.

[33] V. V. Ryazanov, V. V. Bol'ginov, D. S. Sobanin, I. V. Vernik, S. K. Tolpygo, A. M. Kadin, and O. A. Mukhanov. Magnetic Josephson Junction Technology for Digital and Memory Applications. *Physics Procedia*, 36(0):35–41, 2012.

[34] L. N. Bulaevskiĭ, V. V. Kuziĭ, and A. A. Sobyanin. Superconducting system with weak coupling to the current in the ground state. *Soviet Journal of Experimental and Theoretical Physics Letters*, 25:290, April 1977.

[35] W. C. Stewart. Current-Voltage Characteristics of Josephson Junctions. *Applied Physics Letters*, 12(8):277–280, 1968.

[36] D. E. McCumber. Effect of ac Impedance on dc Voltage-Current Characteristics of Superconductor Weak-Link Junctions. *Journal of Applied Physics*, 39(7):3113–3118, 1968.

[37] W. C. Scott. Hyteresis in the dc Switching Characteristics of Josephson Junctions. *Applied Physics Letters*, 17(4):166–169, 1970.

[38] V. Ambegaokar and A. Baratoff. Tunneling Between Superconductors. *Phys. Rev. Lett.*, 10:486–489, June 1963.

[39] V. Ambegaokar and A. Baratoff. Tunneling Between Superconductors; errata. *Phys. Rev. Lett.*, 11:104–104, July 1963.

[40] T. P. Sheahen. Rules for the Energy Gap and Critical Field of Superconductors. *Phys. Rev.*, 149:368–370, Sep 1966.

[41] J. M. Rowell, M. Gurvitch, and J. Geerk. Modification of tunneling barriers on Nb by a few monolayers of Al. *Phys. Rev. B*, 24:2278–2281, Aug 1981.

[42] M. Gurvitch, J. M. Rowell, H. A. Huggins, M. A. Washington, and T. A. Fulton. Nb Josephson tunnel junctions with thin layers of Al near the barrier. In *Electron Devices Meeting, 1981 International*, volume 27, pages 115 – 117, 1981.

[43] M. Gurvitch, M. A. Washington, and H. A. Huggins. High quality refractory Josephson tunnel junctions utilizing thin aluminum layers. *Applied Physics Letters*, 42(5):472–474, 1983.

[44] R. A. Ferrell. Josephson Tunneling and Quantum Mechanical Phase. *Phys. Rev. Lett.*, 15:527–529, Sep 1965.

[45] M. Weihnacht. Influence of Film Thickness on D. C. Josephson Current. *physica status solidi (b)*, 32(2):K169–K172, 1969.

[46] H. A. Kramers. Brownian motion in a field of force and the diffusion model of chemical reactions. *Physica*, 7(4):284 – 304, 1940.

[47] C. Kaiser. *High quality Nb/Al-AlO$_x$/Nb Josephson junctions*. KIT Scientific Publishing, 2011.

[48] T. A. Fulton and L. N. Dunkleberger. Lifetime of the zero-voltage state in Josephson tunnel junctions. *Phys. Rev. B*, 9:4760–4768, Jun 1974.

[49] J. Kurkijärvi. Intrinsic Fluctuations in a Superconducting Ring Closed with a Josephson Junction. *Phys. Rev. B*, 6:832–835, Aug 1972.

[50] L. D. Jackel, W. W. Webb, J. E. Lukens, and S. S. Pei. Measurement of the probability distribution of thermally excited fluxoid quantum transitions in a superconducting ring closed by a Josephson junction. *Phys. Rev. B*, 9:115–118, Jan 1974.

[51] A. Garg. Escape-field distribution for escape from a metastable potential well subject to a steadily increasing bias field. *Phys. Rev. B*, 51:15592–15595, Jun 1995.

[52] M. H. Devoret, J. M. Martinis, D. Esteve, and J. Clarke. Resonant Activation from the Zero-Voltage State of a Current-Biased Josephson Junction. *Phys. Rev. Lett.*, 53:1260–1263, Sep 1984.

[53] A. O. Caldeira and A. J. Leggett. Influence of Dissipation on Quantum Tunneling in Macroscopic Systems. *Phys. Rev. Lett.*, 46:211–214, Jan 1981.

[54] H. Grabert and U. Weiss. Crossover from Thermal Hopping to Quantum Tunneling. *Phys. Rev. Lett.*, 53:1787–1790, Nov 1984.

[55] A. Wallraff, A. Lukashenko, C. Coqui, A. Kemp, T. Duty, and A. V. Ustinov. Switching current measurements of large area Josephson tunnel junctions. *Review of Scientific Instruments*, 74(8):3740–3748, 2003.

[56] A. C. Scott. Distributed device applications of the superconducting tunnel junction. *Solid-State Electronics*, 7(2):137 – 146, 1964.

[57] J. C. Swihart. Field Solution for a Thin-Film Superconducting Strip Transmission Line. *Journal of Applied Physics*, 32(3):461–469, 1961.

[58] R. K. Bullough and P. J. Caudrey. *Solitons*. Springer-Verlag Berlin Heidelberg New York, 1980.

[59] A. Barone and G. Paterno. *Physics and Applications of the Josephson Effect*, volume 1. John Wiley & Sons, Inc., 1982.

[60] A. V. Ustinov. Solitons in Josephson junctions. *Physica D: Nonlinear Phenomena*, 123(1–4):315–329, 1998. Annual International Conference of the Center for Nonlinear Studies.

[61] A. Wallraff, A. Lukashenko, A. Lisenfeld, J. Kemp, M. V. Fistul, Y. Koval, and A. V. Ustinov. Quantum dynamics of a single vortex. *Nature*, 425:155–158, 2003.

[62] J. Nagel, D. Speer, T. Gaber, A. Sterck, R. Eichhorn, P. Reimann, K. Ilin, M. Siegel, D. Koelle, and R. Kleiner. Observation of Negative Absolute Resistance in a Josephson Junction. *Phys. Rev. Lett.*, 100(21):217001, May 2008.

[63] D. W. McLaughlin and A. C. Scott. Perturbation analysis of fluxon dynamics. *Phys. Rev. A*, 18:1652–1680, Oct 1978.

[64] M. D. Fiske. Temperature and Magnetic Field Dependences of the Josephson Tunneling Current. *Rev. Mod. Phys.*, 36:221–222, Jan 1964.

[65] D. D. Coon and M. D. Fiske. Josephson ac and Step Structure in the Supercurrent Tunneling Characteristic. *Phys. Rev.*, 138:A744–A746, May 1965.

[66] E. Goldobin, A. Sterck, T. Gaber, D. Koelle, and R. Kleiner. Dynamics of Semifluxons in Nb Long Josephson $0 - \pi$ Junctions. *Phys. Rev. Lett.*, 92:057005, Feb 2004.

[67] W. H. Henkels. Accurate measurement of small inductances or penetration depths in superconductors. *Applied Physics Letters*, 32(12):829–831, 1978.

[68] W. H. Henkels. Self-contained measurement of thin-film superconducting penetration depths and nonsuperconducting film thicknesses in Josephson integrated circuits. *Journal of Applied Physics*, 57(3):855–860, 1985.

[69] H. Susanto, E. Goldobin, D. Koelle, R. Kleiner, and S. A. van Gils. Controllable plasma energy bands in a one-dimensional crystal of fractional Josephson vortices. *Phys. Rev. B*, 71:174510, May 2005.

[70] R. C. Jaklevic, J. Lambe, A. H. Silver, and J. E. Mercereau. Quantum Interference Effects in Josephson Tunneling. *Phys. Rev. Lett.*, 12:159–160, Feb 1964.

[71] T. A. Fulton, L. N. Dunkleberger, and R. C. Dynes. Quantum Interference Properties of Double Josephson Junctions. *Phys. Rev. B*, 6:855–875, Aug 1972.

[72] W.-T. Tsang and T. Van Duzer. Influence of the current-phase relation on the critical-current-applied-magnetic-flux dependence in parallel-connected Josephson junctions. *Journal of Applied Physics*, 47(6):2656–2661, 1976.

[73] V. V. Ryazanov, V. A. Oboznov, A. Yu. Rusanov, A. V. Veretennikov, A. A. Golubov, and J. Aarts. Coupling of Two Superconductors through a Ferromagnet: Evidence for a π Junction. *Phys. Rev. Lett.*, 86:2427–2430, Mar 2001.

[74] T. Kontos, M. Aprili, J. Lesueur, F. Genêt, B. Stephanidis, and R. Boursier. Josephson Junction through a Thin Ferromagnetic Layer: Negative Coupling. *Phys. Rev. Lett.*, 89:137007, Sep 2002.

[75] H. Hilgenkamp, Ariando, H.-J. H. Smilde, D. H. A. Blank, G. Rijnders, H. Rogalla, J. R. Kirtley, and C. C. Tsuei. Ordering and manipulation of the magnetic moments in large-scale superconducting π-loop arrays. *Nature*, 422:50–53, Mar 2003.

[76] H. J. H. Smilde, Ariando, D. H. A. Blank, G. J. Gerritsma, H. Hilgenkamp, and H. Rogalla. *d*-Wave-Induced Josephson Current Counterflow in $YBa_2Cu_3O_7$/Nb Zigzag Junctions. *Phys. Rev. Lett.*, 88:057004, Jan 2002.

[77] C. C. Tsuei and J. R. Kirtley. Pairing symmetry in cuprate superconductors. *Rev. Mod. Phys.*, 72:969–1016, Oct 2000.

[78] J. H. Xu, J. H. Miller, and C. S. Ting. π-vortex state in a long 0-π Josephson junction. *Phys. Rev. B*, 51:11958–11961, May 1995.

[79] J. R. Kirtley, C. C. Tsuei, and K. A. Moler. Temperature Dependence of the Half-Integer Magnetic Flux Quantum. *Science*, 285(5432):1373–1375, 1999.

[80] J. R. Kirtley, C. C. Tsuei, Martin Rupp, J. Z. Sun, Lock See Yu-Jahnes, A. Gupta, M. B. Ketchen, K. A. Moler, and M. Bhushan. Direct Imaging of Integer and Half-Integer Josephson Vortices in High-T_c Grain Boundaries. *Phys. Rev. Lett.*, 76:1336–1339, Feb 1996.

[81] E. Goldobin, N. Stefanakis, D. Koelle, and R. Kleiner. Fluxon-semifluxon interaction in an annular long Josephson 0-π junction. *Phys. Rev. B*, 70:094520, Sep 2004.

[82] E. Goldobin, D. Koelle, and R. Kleiner. Ground states and bias-current-induced rearrangement of semifluxons in 0-π long Josephson junctions. *Phys. Rev. B*, 67:224515, Jun 2003.

[83] T. Gaber, E. Goldobin, A. Sterck, R. Kleiner, D. Koelle, M. Siegel, and M. Neuhaus. Nonideal artificial phase discontinuity in long Josephson 0-κ junctions. *Phys. Rev. B*, 72(5):054522, Aug 2005.

[84] K. Vogel, W. P. Schleich, T. Kato, D. Koelle, R. Kleiner, and E. Goldobin. Theory of fractional vortex escape in a long Josephson junction. *Phys. Rev. B*, 80:134515, Oct 2009.

[85] K. Buckenmaier. *Aktivierungsenergie fraktionaler Flusswirbel und Spektroskopie an Vortex-Molekülen in langen Josephsonkontakten*. Dissertation, Universität Tübingen, 2011.

[86] B. A. Malomed and A. V. Ustinov. Creation of classical and quantum fluxons by a current dipole in a long Josephson junction. *Phys. Rev. B*, 69:064502, Feb 2004.

[87] C. Nappi, M. P. Lissitski, and R. Cristiano. Fraunhofer critical-current diffraction pattern in annular Josephson junctions with injected current. *Phys. Rev. B*, 65:132516, Mar 2002.

[88] H. Sickinger. *Messung der Eigenfrequenzen fraktionaler Vortex-Moleküle in annularen langen Josephson-Kontakten*. Master thesis, Universität Tübingen, 2009.

[89] A. M. Barychev. *Superconductor-Insulator-Superconductor THz Mixer Integrated with a Superconductin Flux-Flow Oscillator*. Phd thesis, Delft University of Technology, Delft; Russian Academy of Science, Moscow; Space Research Organization of the Netherlands; Dutch Research School for Astronomy, 2005.

[90] D. D. Arnone, C. M. Ciesla, A. Corchia, S. Egusa, M. Pepper, J. M. Chamberlain, C. Bezant, E. H. Linfield, R. Clothier, and N. Khammo. Applications of terahertz (THz) technology to medical imaging. In *Industrial Lasers and Inspection (EUROPTO Series)*, pages 209–219. International Society for Optics and Photonics, 1999.

[91] T. May, G. Zieger, S. Anders, V. Zakosarenko, M. Starkloff, H. G. Meyer, G. Thorwirth, and E. Kreysa. Passive stand-off terahertz imaging with 1 hertz frame rate. In *SPIE Defense and Security Symposium*, pages 69490C–69490C. International Society for Optics and Photonics, 2008.

[92] J. R. Tucker and M. J. Feldman. Quantum detection at millimeter wavelengths. *Rev. Mod. Phys.*, 57:1055–1113, Oct 1985.

[93] P. H. Siegel. Terahertz technology. *Microwave Theory and Techniques, IEEE Transactions on*, 50(3):910 –928, mar 2002.

[94] K. S. Il'in, M. Lindgren, M. Currie, A. D. Semenov, G. N. Gol'tsman, Roman Sobolewski, S. I. Cherednichenko, and E. M. Gershenzon. Picosecond hot-electron energy relaxation in NbN superconducting photodetectors. *Applied Physics Letters*, 76(19):2752–2754, 2000.

[95] A. D. Semenov, G. N. Gol'tsman, and R. Sobolewski. Hot-electron effect in superconductors and its applications for radiation sensors. *Superconductor Science and Technology*, 15(4):R1, 2002.

[96] G. J. Dolan, T. G. Phillips, and D. P. Woody. Low-noise 115-GHz mixing in superconducting oxide-barrier tunnel junctions. *Applied Physics Letters*, 34(5):347–349, 1979.

[97] P. L. Richards, T. M. Shen, R. E. Harris, and F. L. Lloyd. Quasiparticle heterodyne mixing in SIS tunnel junctions. *Applied Physics Letters*, 34(5):345–347, 1979.

[98] S. Rudner and T. Claeson. Arrays of superconducting tunnel junctions as low-noise 10-GHz mixers. *Applied Physics Letters*, 34(10):711–713, 1979.

[99] J. Qin, K. Enpuku, and K. Yoshida. Flux-flow-type Josephson oscillator for millimeter and submillimeter wave region. IV. Thin-film coupling. *Journal of Applied Physics*, 63(4):1130–1135, 1988.

[100] M. Yu. Torgashin, V. P. Koshelets, P. N. Dmitriev, A. B. Ermakov, L. V. Filippenko, and P. A. Yagoubov. Superconducting Integrated Receiver Based on Nb-AlN-NbN-Nb Circuits. *Applied Superconductivity, IEEE Transactions on*, 17(2):379 –382, june 2007.

[101] J. Villegier, L. Vieux-Rochaz, M. Goniche, P. Renard, and M. Vabre. NbN tunnel junctions. *Magnetics, IEEE Transactions on*, 21(2):498 – 504, mar 1985.

[102] F. Shinoki, A. Shoji, S. Kosaka, S. Takada, and H. Hayakawa. Niobium nitride Josephson tunnel junctions with oxidized amorphous silicon barriers. *Applied Physics Letters*, 38(4):285–286, 1981.

[103] A. Shoji, F. Shinoki, S. Kosaka, M. Aoyagi, and H. Hayakawa. New fabrication process for Josephson tunnel junctions with (niobium nitride, niobium) double-layered electrodes. *Applied Physics Letters*, 41(11):1097–1099, 1982.

[104] H. G. LeDuc, S. K. Khanna, and J. A. Stern. All NbN tunnel junction fabrication. *International Journal of Infrared and Millimeter Waves*, 8:1243–1248, 1987.

[105] P. N. Dmitriev, I. L. Lapitskaya, L. V. Filippenko, A. B. Ermakov, S. V. Shitov, G. V. Prokopenko, S. A. Kovtonyuk, and V. P. Koshelets. High quality Nb-based tunnel junctions for high frequency and digital applications. *Applied Superconductivity, IEEE Transactions on*, 13(2):107–110, 2003.

[106] Herschel website at ESA. http://astro.estec.esa.nl/SAgeneral/Projects/First/first.html.

[107] Th. de Graauw, F. P. Helmich, T. G. Phillips, J. Stutzki, E. Caux, N. D. Whyborn, P. Dieleman, P. R. Roelfsema, H. Aarts, R. Assendorp, R. Bachiller, W. Baechtold, A. Barcia, D. A. Beintema, V. Belitsky, A. O. Benz, R. Bieber, A. Boogert, C. Borys, B. Bumble, P. Caïs, M. Caris, P. Cerulli-Irelli, G. Chattopadhyay, S. Cherednichenko, M. Ciechanowicz, O. Coeur-Joly, C. Comito, A. Cros, A. de Jonge, G. de Lange, B. Delforges, Y. Delorme, T. den Boggende, J.-M. Desbat, C. Diez-González, A. M. Di Giorgio, L. Dubbeldam, K. Edwards, M. Eggens, N. Erickson, J. Evers, M. Fich, T. Finn, B. Franke, T. Gaier, C. Gal, J. R. Gao, J.-D. Gallego, S. Gauffre, J. J. Gill, S. Glenz, H. Golstein, H. Goulooze, T. Gunsing, R. Güsten, P. Hartogh, W. A. Hatch, R. Higgins, E. C. Honingh, R. Huisman, B. D. Jackson, H. Jacobs, K. Jacobs, C. Jarchow, H. Javadi, W. Jellema, M. Justen, A. Karpov, C. Kasemann, J. Kawamura, G. Keizer, D. Kester, T. M. Klapwijk, Th. Klein, E. Kollberg, J. Kooi, P.-P. Kooiman, B. Kopf, M. Krause, J.-M. Krieg, C. Kramer, B. Kruizenga, T. Kuhn, W. Laauwen, R. Lai, B. Larsson, H. G. Leduc, C. Leinz, R. H. Lin, R. Liseau, G. S. Liu, A. Loose, I. López-Fernandez, S. Lord, W. Luinge, A. Marston, J. Martín-Pintado, A. Maestrini, F. W. Maiwald, C. McCoey, I. Mehdi, A. Megej, M. Melchior, L. Meinsma, H. Merkel, M. Michalska, C. Monstein, D. Moratschke, P. Morris, H. Muller, J. A. Murphy, A. Naber, E. Natale, W. Nowosielski, F. Nuzzolo, M. Olberg, M. Olbrich, R. Orfei, P. Orleanski, V. Ossenkopf, T. Peacock, J. C. Pearson, I. Peron, S. Phillip-May, L. Piazzo, P. Planesas, M. Rataj, L. Ravera, C. Risacher, M. Salez, L. A. Samoska, P. Saraceno, R. Schieder, E. Schlecht, F. Schlöder, F. Schmülling, M. Schultz, K. Schuster, O. Siebertz, H. Smit, R. Szczerba, R. Shipman, E. Steinmetz, J. A. Stern, M. Stokroos, R. Teipen, D. Teyssier, T. Tils, N. Trappe, C. van Baaren, B.-J. van Leeuwen, H. van de Stadt, H. Visser, K. J. Wildeman, C. K. Wafelbakker, J. S. Ward, P. Wesselius, W. Wild, S. Wulff, H.-J. Wunsch, X. Tielens, P. Zaal, H. Zirath, J. Zmuidzinas, and F. Zwart. The Herschel-Heterodyne Instrument for the Far-Infrared (HIFI)*. *A&A*, 518:L6, 2010.

[108] Website of the ALMA project. https://almascience.nrao.edu; Atacama Large Millimeter/submillimeter Array (ALMA).

[109] Manfred Birk, Georg Wagner, Gert de Lange, Arno de Lange, Brian N Ellison, Mark R Harman, Axel Murk, Hermann Oelhaf, Guido Maucher, and Christian Sartorius. TELIS: TErahertz and subMMW LImb Sounder–Project Summary After First Successful Flight. In *21st international symposium on space terahertz technology*, 2010.

[110] Website of Oerlikon Leybold Vacuum. http://www.oerlikon.com/leyboldvacuum/de.

[111] J. Hinken. *Supraleiter-Elektronik*. Springer-Verlag, Berlin, 1988.

[112] T. Van Duzer and C. W. Turner. *Principles of superconductive devices and circuits*, volume 31. Elsevier New York, 1981.

[113] W. Henkels and C. Kircher. Penetration depth measurements on type ii superconducting films. *Magnetics, IEEE Transactions on*, 13(1):63 – 66, Jan 1977.

[114] T. Imamura and S. Hasuo. Cross-sectional transmission electron microscopy observation of Nb/AlO_x-Al/Nb Josephson junctions. *Applied Physics Letters*, 58(6):645–647, 1991.

[115] T. Imamura and S. Hasuo. Fabrication of high quality Nb/AlO_x-Al/Nb Josephson junctions. II. Deposition of thin Al layers on Nb films. *Applied Superconductivity, IEEE Transactions on*, 2(2):84 –94, Jun 1992.

[116] A. A. Golubov, E. P. Houwman, J. G. Gijsbertsen, V. M. Krasnov, J. Flokstra, H. Rogalla, and M. Yu. Kupriyanov. Proximity effect in superconductor-insulator-superconductor Josephson tunnel junctions: Theory and experiment. *Phys. Rev. B*, 51:1073–1089, Jan 1995.

[117] J. Antes. *Entwurf und Konstruktion einer Steuerung für die kontrollierte Oxidation von Aluminium*. Studienarbeit, Universität Karlsruhe, 2009.

[118] R. E. Miller, W. H. Mallison, A. W. Kleinsasser, K. A. Delin, and E. M. Macedo. Niobium trilayer Josephson tunnel junctions with ultrahigh critical current densities. *Applied Physics Letters*, 63(10):1423–1425, 1993.

[119] H. Sugiyama, A. Fujimaki, and H. Hayakawa. Characteristics of high critical current density Josephson junctions with Nb/AlOx/Nb trilayers. *Applied Superconductivity, IEEE Transactions on*, 5(2):2739 – 2742, Jun 1995.

[120] K. Il'in, M. Siegel, A. Semenov, A. Engel, and H.-W. Hübers. Critical current of Nb and NbN thin-film structures: The cross-section dependence. *physica status solidi (c)*, 2(5):1680–1687, 2005.

[121] K. S. Il'in, D. Rall, M. Siegel, and A. Semenov. Critical current density in thin superconducting TaN film structures. *Physica C: Superconductivity*, 479(0):176–178, 2012. Proceedings of VORTEX VII Conference.

[122] ALLRESIST GmbH. http://www.allresist.de; Allresist is a distributor of various lithography products, including EBL resists and developer.

[123] J. M. Meckbach. *Entwicklung langer Josephson-Kontakte mit sub-μm Strominjektoren*. Diploma thesis, Universität Karlsruhe, 2009.

[124] Internal report from Dr. rer. nat. Tobias Gaber, Physikalisches Institut Experimentalphysik II (PIT II), Universität Tübingen, Germany.

[125] T. Gaber. *Measurements at mK*, 2011. Presentation given at the "Fluxon Dynamics Seminar 2011 KIT - IMS".

[126] R. F. Broom, A. Oosenbrug, and W. Walter. Josephson junctions of small area formed on the edges of niobium films. *Applied Physics Letters*, 37(2):237–239, 1980.

[127] H. Yamamori, T. Yamada, H. Sasaki, and A. Shoji. A 10 V programmable Josephson voltage standard circuit with a maximum output voltage of 20 V. *Superconductor Science and Technology*, 21(10):105007, 2008.

[128] S. Anders, M. Schmelz, L. Fritzsch, R. Stolz, V. Zakosarenko, T. Schönau, and H.-G. Meyer. Sub-micrometer-sized, cross-type Nb-AlO$_x$-Nb tunnel junctions with low parasitic capacitance. *Superconductor Science and Technology*, 22(6):064012, 2009.

[129] L. V. Filippenko, S. V. Shitov, P. N. Dmitriev, A. B. Ermakov, V. P. Koshelets, and J.-R. Gao. Submillimeter superconducting integrated receivers: Fabrication and yield. *Applied Superconductivity, IEEE Transactions on*, 11(1):816 –819, mar 2001.

[130] K. Kuroda and M. Yuda. Niobium-stress influence on Nb/Al-oxide/Nb Josephson junctions. *Journal of Applied Physics*, 63(7):2352–2357, 1988.

[131] S. Jauß and G. Göring. *Charakterisierung von Palladium und Manganin auf unterschiedlichen Substraten*, 2011. "Nanopraktikum" at IMS- KIT; Supervisor: J. M. Meckbach.

[132] Data aquisition program for measurment of Josephson junctions. The program was written by Dr. Edward Goldobin. http://www.geocities.com/goldexi.

[133] M. Kusunoki, H. Yamamori, A. Fujimaki, Y. Takai, and H. Hayakawa. High-Quality Nb/AlO$_x$-Al/Nb Josephson Junctions with Gap Voltage of 2.95 mV. *Japanese Journal of Applied Physics*, 32(Part 2, No. 11A):L1609–L1611, 1993.

[134] N. Bolse and A. Lochbaum. *Untersuchung von SiO$_2$ zur Herstellung von Josephson-Kontakten*, 2010. "Nanopraktikum" at IMS- KIT; Supervisor: J. M. Meckbach.

[135] J.-G. Caputo, N. Flytzanis, and E. Vavalis. Effect of geometry on fluxon width in a Josephson junction. *International Journal of Modern Physics C*, 07(02):191–216, 1996.

[136] A. Franz, A. Wallraff, and A. V. Ustinov. Magnetic field penetration in a long Josephson junction imbedded in a wide stripline. *Journal of Applied Physics*, 89(1):471–476, 2001.

[137] M. R. Samuelsen and S. A. Vasenko. Influence of the bias current distribution on the static and dynamic properties of long Josephson junctions. *Journal of Applied Physics*, 57(1):110–112, 1985.

[138] S. Pagano, B. Ruggiero, and E. Sarnelli. Magnetic-field dependence of the critical current in long Josephson junctions. *Phys. Rev. B*, 43:5364–5369, Mar 1991.

[139] M. Rudolph. *Asymmetrische dc SQUIDs*. Diploma thesis, Universität Tübingen, 2012.

[140] J. Nagel. *Asymmetrische SQUIDs und nanoSQUIDs: Quantuminterferometer unter neuartigen Bedingungen*. Dissertation, Universität Tübingen, 2012.

[141] STAR Cryoelectronics. http://www.starcryo.com; Star Cryoelectronics manufactures Superconducting Quantum Interference Device (SQUID) sensors and advanced PC-based SQUID control electronics.

[142] Private communication with Prof. Dr. rer. nat. Reinhold Kleiner, Physikalisches Institut Experimentalphysik II (PIT II), Universität Tübingen, Germany.

[143] T. Schönau, M. Schmelz, V. Zakosarenko, R. Stolz, S. Anders, L. Fritzsch, and H.-G. Meyer. SQIF-based dc SQUID amplifier with intrinsic negative feedback. *Superconductor Science and Technology*, 25(1):015005, 2012.

[144] N. Martucciello, J. Mygind, V. P. Koshelets, A. V. Shchukin, L. V. Filippenko, and Roberto Monaco. Fluxon dynamics in long annular Josephson tunnel junctions. *Phys. Rev. B*, 57:5444–5449, Mar 1998.

[145] T. Gaber. *Dynamik fraktionaler Flusswirbelin langen Josephsonkontakten*. Dissertation, Universität Tübingen, 2007.

[146] K. Buckenmaier, T. Gaber, M. Siegel, D. Koelle, R. Kleiner, and E. Goldobin. Spectroscopy of the Fractional Vortex Eigenfrequency in a Long Josephson 0-κ Junction. *Phys. Rev. Lett.*, 98(11):117006, Mar 2007.

[147] T. Gaber, K. Buckenmaier, D. Koelle, R. Kleiner, and E. Goldobin. Fractional Josephson vortices: oscillating macroscopic spins. *Applied Physics A*, 89(3):587–592, 2007.

[148] U. Kienzle, T. Gaber, K. Buckenmaier, K. Ilin, M. Siegel, D. Koelle, R. Kleiner, and E. Goldobin. Thermal escape of fractional vortices in long Josephson junctions. *Phys. Rev. B*, 80(1):014504, Jul 2009.

[149] E. Goldobin, H. Susanto, D. Koelle, R. Kleiner, and S. A. van Gils. Oscillatory eigenmodes and stability of one and two arbitrary fractional vortices in long Josephson 0-κ junctions. *Phys. Rev. B*, 71:104518, Mar 2005.

[150] StkJJ (Stacked Josephson Junction simulator) allows to make numerical simulations of long Josephson junctions. StkJJ essentially solves (a system of) perturbed sine-Gordon equation(s) that define the dynamics of the Josephson phase in the junction(s). http://www.geocities.com/SiliconValley/Heights/7318/StkJJ.htm.

[151] E. Goldobin, D. Koelle, and R. Kleiner. Ground states of one and two fractional vortices in long Josephson 0-κ junctions. *Phys. Rev. B*, 70:174519, Nov 2004.

[152] A. Dewes, T. Gaber, D. Koelle, R. Kleiner, and E. Goldobin. Semifluxon Molecule under Control. *Phys. Rev. Lett.*, 101:247001, Dec 2008.

[153] T. Kato and M. Imada. Vortices and Quantum Tunneling in Current-Biased 0-π-0 Josephson Junctions of d-Wave Superconductors. *Journal of the Physical Society of Japan*, 66(5):1445–1449, 1997.

[154] E. Goldobin, K. Vogel, O. Crasser, R. Walser, W. P. Schleich, D. Koelle, and R. Kleiner. Quantum tunneling of semifluxons in a 0-π-0 long Josephson junction. *Phys. Rev. B*, 72:054527, Aug 2005.

[155] D. M. Heim, K. Vogel, W. P. Schleich, D. Koelle, R. Kleiner, and E. Goldobin. A tunable macroscopic quantum system based on two fractional vortices. *New Journal of Physics*, 15(5):053020, 2013.

[156] H. Dalsgaard Jensen, A. Larsen, and J. Mygind. Resonator coupled Josephson junctions; parametric excitations and mutual locking. *Magnetics, IEEE Transactions on*, 27(2):3355–3358, 1991.

[157] A. Benabdallah, J. G. Caputo, and A. C. Scott. Exponentially tapered Josephson flux-flow oscillator. *Physical Review B*, 54(22):16139, 1996.

[158] Private communication with Dr. A. Sobolev, Institute of Radioengineering and Electronics (IRE) of Russian Academy of Sciences (RAS), Moscow, Russia.

[159] A. Shoji, M. Aoyagi, S. Kosaka, F. Shinoki, and H. Hayakawa. Niobium nitride Josephson tunnel junctions with magnesium oxide barriers. *Applied Physics Letters*, 46(11):1098–1100, 1985.

[160] J. R. Kirtley, S. I. Raider, R. M. Feenstra, and A. P. Fein. Spatial variation of the observed energy gap in granular superconducting NbN films. *Applied Physics Letters*, 50(22):1607–1609, 1987.

[161] Z. Wang, A. Kawakami, Y. Uzawa, and B. Komiyama. Superconducting properties and crystal structures of single-crystal niobium nitride thin films deposited at ambient substrate temperature. *Journal of Applied Physics*, 79(10):7837–7842, 1996.

[162] A. Kawakami, Z. Wang, and S. Miki. Fabrication and characterization of epitaxial NbN/MgO/NbN Josephson tunnel junctions. *Journal of Applied Physics*, 90(9):4796–4799, 2001.

[163] B. Bumble, H. G. Leduc, J. A. Stern, and K. G. Megerian. Fabrication of Nb/AlNx/NbTiN junctions for SIS mixer applications. *Applied Superconductivity, IEEE Transactions on*, 11(1):76 –79, Mar 2001.

[164] R. H. French. Electronic Band Structure of Al2O3, with Comparison to Alon and AlN. *Journal of the American Ceramic Society*, 73(3):477–489, 1990.

[165] W. M. Yim, E. J. Stofko, P. J. Zanzucchi, J. I. Pankove, M. Ettenberg, and S. L. Gilbert. Epitaxially grown AlN and its optical band gap. *Journal of Applied Physics*, 44(1):292–296, 1973.

[166] T. Shiino, S. Shiba, N. Sakai, T. Yamakura, L. Jiang, Y. Uzawa, H. Maezawa, and S. Yamamoto. Improvement of the critical temperature of superconducting NbTiN and NbN thin films using the AlN buffer layer. *Superconductor Science and Technology*, 23(4):045004, 2010.

[167] Z. Wang, A. Kawakami, Y. Uzawa, and B. Komiyama. High critical current density NbN/AlN/NbN tunnel junctions fabricated on ambient temperature MgO substrates. *Applied Physics Letters*, 64(15):2034–2036, 1994.

[168] J. A. Stern. *Fabrication and testing of NbN/MgO/NbN tunnel junctions for use as gigh-frequency heterodyne detectors*. PhD thesis, California Institute of Technology, Pasedena, California, USA, 1991.

[169] M. Aoyagi, H. Nakagawa, I. Kurosawa, and S. Takada. NbN/MgO/NbN Josephson junctions for integrated circuits. *Japanese Journal of Applied Physics*, 31(part 1):1778–1783, 1992.

[170] Z. Wang, A. Kawakami, and Y. Uzawa. Nbn/aln/nbn tunnel junctions with high current density up to 54 ka/cm[sup 2]. *Applied Physics Letters*, 70(1):114–116, 1997.

[171] Z. Wang, H. Terai, A. Kawakami, and Y. Uzawa. Interface and tunneling barrier heights of NbN/AlN/NbN tunnel junctions. *Applied Physics Letters*, 75(5):701–703, 1999.

[172] Z. Wang, H. Terai, A. Kawakami, and Y. Uzawa. Characterization of NbN/AlN/NbN tunnel junctions. *Applied Superconductivity, IEEE Transactions on*, 9(2):3259–3262, jun 1999.

[173] Y. Nakamura, H. Terai, K. Inomata, T. Yamamoto, W. Qiu, and Z. Wang. Superconducting qubits consisting of epitaxially grown NbN/AlN/NbN Josephson junctions. *Applied Physics Letters*, 99(21):212502, 2011.

[174] H. Terai and Wang Z. All-NbN single flux quantum circuits based on NbN/AlN/NbN tunnel junctions. *IEICE Transactions on Electronics*, 83(1):69–74, 2000.

[175] H. Terai, A. Kawakami, and Z. Wang. Sub-micron NbN/AlN/NbN tunnel junction with high critical current density. *Physica C: Superconductivity*, 372(0):38–41, 2002.

[176] Website of Delta Elektronika B.V. http://www.delta-elektronika.nl/en.

[177] Website of Invensys Operations Management Eurotherm. http://www.eurotherm.de; Invensys Operations Managemnet.

[178] Website of Kurt J. Lesker Company. http://lesker.com/.

[179] Website of Advanced Energy. http://www.advanced-energy.de.

[180] B. Chapman. *Glow discharge processes: sputtering and plasma etching*. Wiley New York, 1980.

[181] S. Kadlec, J. Musil, and H. Vyskocil. Hysteresis effect in reactive sputtering: a problem of system stability. *Journal of Physics D: Applied Physics*, 19(9):L187, 1986.

[182] G. Zinsmeister. Theory of thin film condensation. Part B: Solution of the simplified condensation equation. *Thin Solid Films*, 2(5):497–507, 1968.

[183] N. N. Iosad, T. M. Klapwijk, S. N. Polyakov, V. V. Roddatis, E. K. Kov'ev, and P. N. Dmitriev. Properties of DC magnetron sputtered Nb and NbN films for different source conditions. *Applied Superconductivity, IEEE Transactions on*, 9(2):1720–1723, 1999.

[184] K. Il'in, M. Siegel, A. Engel, H. Bartolf, A. Schilling, A. Semenov, and H.-W. Hue-bers. Current-Induced Critical State in NbN Thin-Film Structures. *Journal of Low Temperature Physics*, 151:585–590, 2008.

[185] A. Semenov, B. Günther, U. Böttger, H.-W. Hübers, H. Bartolf, A. Engel, A. Schilling, K. Ilin, M. Siegel, R. Schneider, D. Gerthsen, and N. A. Gippius. Op-tical and transport properties of ultrathin NbN films and nanostructures. *Physical Review B*, 80(5):054510, 2009.

[186] S. Thakoor, J. L. Lamb, A. P. Thakoor, and S. K. Khanna. High T_c superconducting NbN films deposited at room temperature. *Journal of Applied Physics*, 58(12):4643–4648, 1985.

[187] Hiroyuki Akaike, Tatsunori Funai, Shunya Skamoto, and Akira Fujimaki. Fabrica-tion of NbN/Al-AlN$_x$/NbN Tunnel Junctions on Several Kinds of Substrates. In *14th International Superconductive Electronics Conference*, 2013.

[188] K. L. Westra, M. J. Brett, and J. F. Vaneldik. Properties of reactively sputtered NbN films. *Journal of Vacuum Science & Technology A: Vacuum, Surfaces, and Films*, 8(3):1288–1293, 1990.

[189] S. H. Bedorf. *Development of Ultrathin Niobium Nitride and Niobium Titanium Nitride Films for THz Hot-Electron Bolometers*. Dissertation, Universität zu Köln, 2005.

[190] B. J. Feenstra, F. C. Klaassen, D. van der Marel, Z. H. Barber, R. Perez Pinaya, and M. Decroux. Penetration depth and conductivity of NbN and DyBa$_2$Cu$_3$O$_{7-\delta}$ thin films measured by mm-wave transmission. *Physica C: Superconductivity*, 278(3):213–222, 1997.

[191] B. Komiyama, Z. Wang, and M. Tonouchi. Penetration depth measurements of single-crystal NbN films at millimeter-wave region. *Applied Physics Letters*, 68(4):562–563, 1996.

[192] S. Kubo, M. Asahi, M. Hikita, and M. Igarashi. Magnetic penetration depths in superconducting NbN films prepared by reactive dc magnetron sputtering. *Applied Physics Letters*, 44(2):258–260, 1984.

[193] M. S. Pambianchi, S. M. Anlage, E. S. Hellman, E. H. Hartford, M. Bruns, and S. Y. Lee. Penetration depth, microwave surface resistance, and gap ratio in NbN and Ba$_{1-x}$K$_x$BiO$_3$ thin films. *Applied Physics Letters*, 64(2):244–246, 1994.

[194] T. E. Takken, M. R. Beasley, and R. F. W. Pease. Penetration depth and critical current in NbN resonators: predicting nonlinearities and breakdown in microstrip. *Applied Superconductivity, IEEE Transactions on*, 5(2):1975–1978, 1995.

[195] A. Shoji, S. Kiryu, and S. Kohjiro. Superconducting properties and normal-state resistivity of single-crystal NbN films prepared by a reactive rf-magnetron sputtering method. *Applied Physics Letters*, 60(13):1624–1626, 1992.

[196] J. M. Murduck, J. Porter, W. Dozier, R. Sandell, J. Burch, J. Bulman, C. Dang, L. Lee, H. Chan, R. W. Simon, and A. H. Silver. Niobium trilayer process for superconducting circuits. *Magnetics, IEEE Transactions on*, 25(2):1139–1142, 1989.

[197] T. P. Orlando, E. J. McNiff, S. Foner, and M. R. Beasley. Critical fields, Pauli paramagnetic limiting, and material parameters of Nb_3Sn and V_3Si. *Phys. Rev. B*, 19:4545–4561, May 1979.

[198] J. M. Murduck, J. Vicent, Ivan K. Schuller, and J. B. Ketterson. Fabrication of NbN/AlN superconducting multilayers. *Journal of Applied Physics*, 62(10):4216–4219, 1987.

[199] 2010 The NIST reference on constants, units and uncertainty. Fundamental physics constants. http://physics.nist.gov/cuu/Constants.

[200] P. J. Mohr, B. N. Taylor, and D. B. Newell. CODATA recommended values of the fundamental physical constants: 2006. *Rev. Mod. Phys.*, 80:633–730, Jun 2008.

[201] A. S. Cooper. Precise lattice constants of germanium, aluminum, gallium arsenide, uranium, sulphur, quartz and sapphire. *Acta Crystallographica*, 15(6):578–582, Jun 1962.

[202] V. Cimalla, J. Pezoldt, G. Ecke, R. Kosiba, O. Ambacher, L. Spiess, G. Teichert, H. Lu, and W. J. Schaff. Growth of cubic InN on r-plane sapphire. *Applied Physics Letters*, 83(17):3468–3470, 2003.

[203] C. Kittel. *Einführung in die Festkörperphysik*, volume Translation of the 3rd edition. R. Oldenbourg Verlag München, 1968.

[204] Wikimedia Commons Website. http://commons.wikimedia.org.

[205] W. J. Meng, J. Heremans, and Y. T. Cheng. Epitaxial growth of aluminum nitride on Si (111) by reactive sputtering. *Applied Physics Letters*, 59(17):2097–2099, 1991.

[206] R. Liu, F. A. Ponce, A. Dadgar, and A. Krost. Atomic arrangement at the AlN/Si (111) interface. *Applied Physics Letters*, 83(5):860–862, 2003.

[207] I. Ivanov, L. Hultman, K. Jarrendahl, P. Martensson, J.-E. Sundgren, B. Hjorvarsson, and J. E. Greene. Growth of epitaxial AlN (0001) on Si (111) by reactive magnetron sputter deposition. *Journal of Applied Physics*, 78(9):5721–5726, 1995.

[208] J. X. Zhang, H. Cheng, Y. Z. Chen, A. Uddin, S. Yuan, S. J. Geng, and S. Zhang. Growth of AlN films on Si (100) and Si (111) substrates by reactive magnetron sputtering. *Surface and Coatings Technology*, 198(1):68–73, 2005.

[209] L. Spieß, G. Teichert, R. Schwarzer, H. Behnken, and C. Genzel. *Moderne Röntgenbeugung: Röntgendiffraktometrie für Materialwissenschaftler, Physiker und Chemiker*. Springer DE, 2009.

[210] M. A. Lewis, D. A. Glocker, and J. Jorne. Measurements of secondary electron emission in reactive sputtering of aluminum and titanium nitride. *Journal of Vacuum Science & Technology A: Vacuum, Surfaces, and Films*, 7(3):1019–1024, 1989.

[211] I. Petrov, A. Myers, J. E. Greene, and J. R. Abelson. Mass and energy resolved detection of ions and neutral sputtered species incident at the substrate during reactive magnetron sputtering of Ti in mixed Ar+N$_2$ mixtures. *Journal of Vacuum Science & Technology A: Vacuum, Surfaces, and Films*, 12(5):2846–2854, 1994.

[212] Ü. Özgür, G. Webb-Wood, H. O. Everitt, F. Yun, and H. Morkoç. Systematic measurement of Al$_x$Ga$_{1-x}$N refractive indices. *Applied Physics Letters*, 79(25):4103–4105, 2001.

[213] D. Zhuang and J. H. Edgar. Wet etching of GaN, AlN, and SiC: a review. *Materials Science and Engineering: R: Reports*, 48(1):1–46, 2005.

[214] Z. Wang, Y. Uzawa, and A. Kawakami. High current density NbN/AlN/NbN tunnel junctions for submillimeter wave SIS mixers. *Applied Superconductivity, IEEE Transactions on*, 7(2):2797–2800, 1997.

[215] MicroChemicals GmbH. http://www.microchemicals.de; MicroChemicals is a distributor of various resists and developers.

Own Publications

[HRH+12] D. Henrich, P. Reichensperger, M. Hofherr, J. M. Meckbach, K. Il'in, M. Siegel, A. Semenov, A. Zotova, and D. Yu. Vodolazov. Geometry-induced reduction of the critical current in superconducting nanowires. *Phys. Rev. B*, 86:144504, Oct 2012.

[KMB+12] U. Kienzle, J. M. Meckbach, K. Buckenmaier, T. Gaber, H. Sickinger, Ch. Kaiser, K. Ilin, M. Siegel, D. Koelle, R. Kleiner, and E. Goldobin. Spectroscopy of a fractional Josephson vortex molecule. *Phys. Rev. B*, 85:014521, Jan 2012.

[KMI+11] Ch. Kaiser, J. M. Meckbach, K. S. Ilin, J. Lisenfeld, R. Schäfer, A. V. Ustinov, and M. Siegel. Aluminum hard mask technique for the fabrication of high quality submicron Nb/Al-AlO$_x$/Nb Josephson junctions. *Supercond. Sci. Tech.*, 24(3):035005, 2011.

[KSS+13] A.A. Kuzmin, S.V. Shitov, A. Scheuring, J.M. Meckbach, K.S. Il'in, S. Wuensch, A.V. Ustinov, and M. Siegel. TES Bolometers With High-Frequency Readout Circuit. *Terahertz Science and Technology, IEEE Transactions on*, 3(1):25–31, 2013.

[MMB+13] J. M. Meckbach, M. Merker, S. J. Buehler, K. Ilin, B. Neumeier, U. Kienzle, E. Goldobin, R. Kleiner, D. Koelle, and M. Siegel. Sub-μm Josephson Junctions for Superconducting Quantum Devices. *Applied Superconductivity, IEEE Transactions on*, 23(3):1100504, June 2013.

[RNM+12] M. Rudolph, J. Nagel, J. M. Meckbach, M. Kemmler, M. Siegel, K. Ilin, D. Koelle, and R. Kleiner. Direct current superconducting quantum interferometers with asymmetric shunt resistors. *Applied Physics Letters*, 101(5):052602, 2012.

[SAG+12] T. A. Scherer, G. Aiello, G. Grossetti, A. Meier, S. Schreck, P. Spaeh, D. Strauss, A. Vaccaro, M. Siegel, J. M. Meckbach, and A. Scheuring. Reduction of surface losses of CVD diamond by passivation methods. In *Infrared, Millimeter, and Terahertz Waves (IRMMW-THz), 2012 37th International Conference on*, pages 1 –2, Sept. 2012.

[TAY+09] S. K. Tolpygo, D. Amparo, D. T. Yohannes, M. Meckbach, and A. F. Kirichenko. Process-Induced Variability of Nb/Al-AlO$_x$/Nb Junctions in Superconductor Integrated Circuits and Protection Against It. *Applied Superconductivity, IEEE Transactions on*, 19(3):135 –139, June 2009.

[YIM+13] V. V. Yurchenko, K. Ilin, J. M. Meckbach, M. Siegel, A. J. Qviller, Y. M. Galperin, and T. H. Johansen. Thermo-magnetic stability of superconducting films controlled by nano-morphology. *Applied Physics Letters*, 102(25):252601, 2013.

Supervised Student Theses

[Bue11] Buehler, Simon. *Integration einer dritten supraleitenden Metallebene in die Nb-AlO$_x$-Nb Technologie*, 2011. Studienarbeit, Institut für Mikro- und Nanoelektronische Systeme, Karlsruher Institut für Technologie (KIT).

[Bue13] Buehler, Simon. *Creation of artificial 1D vortex crystals using current injectors*, 2013. Diploma thesis, Institut für Mikro- und Nanoelektronische Systeme, Karlsruher Institut für Technologie (KIT).

[Gup11] Gupta, Vijeet. *Development of a NbN/AlN/NbN Multilayer Technology*, 2011. Master thesis, Institut für Mikro- und Nanoelektronische Systeme, Karlsruher Institut für Technologie (KIT).

[Hen10] Heni, Wolfgang. *Untersuchung der Isolationseigenschaften und der dielektrischen Verluste in SiO$_2$*, 2010. Bachelor thesis, Institut für Mikro- und Nanoelektronische Systeme, Karlsruher Institut für Technologie (KIT).

[Heu10] Heunisch, Sebastian. *Charakterisierung der Parameterstreuung von Josephson-Kontakten für die Skalierbarkeit ihres Herstellungsprozesses*, 2010. Bachelor thesis, Institut für Mikro- und Nanoelektronische Systeme, Karlsruher Institut für Technologie (KIT).

[Loc11] Lochbaum, Alexander. *Variable Temperature Characterization of Digital Superconductor Circuits*, 2011. Bachelor thesis, Institut für Mikro- und Nanoelektronische Systeme, Karlsruher Institut für Technologie (KIT) and HYPRES, Inc., 175 Clearbrook Road, Elmsford, NY 10523, USA.

[Mar13] Marko, Pavol. *Optimization of noise temperature of THz HEB mixers based on AlN/NbN multilayers*, 2013. Master thesis, Institut für Mikro- und Nanoelektronische Systeme, Karlsruher Institut für Technologie (KIT).

[Mer12] Merker, Michael. *Mehrstufiger Isolationsprozess zur Planarisierung von Josephson-Kontakten*, 2012. Diploma thesis, Institut für Mikro- und Nanoelektronische Systeme, Karlsruher Institut für Technologie (KIT).

[Rez11] Rezem, Maher. *Optimierung von Ätzprozessen zur Mikro- und Nanostrukturierung*, 2011. Bachelor thesis, Institut für Mikro- und Nanoelektronische Systeme, Karlsruher Institut für Technologie (KIT).

International Conferences

[1] M. Merker, J. M. Meckbach, K. Ilin, and M. Siegel. Planar multilayer process for Josephson junction devices. In *14th International Superconductive Eletronics Conference, Cambridge, MA, USA*, 2013.

[2] J. M. Meckbach, A. Häffelin, S. J. Buehler, M. Merker, K. Ilin, and M. Siegel. Niobium Nitride Technology for Josephson Junction Devices. In *DPG Frühjahrstagung, Regensburg, Germany*, 2013.

[3] M. Merker, J. M. Meckbach, S. Buehler, K. Ilin, and M. Siegel. Self-Planarized Process for the Fabrication of Josephson Junction Devices. In *DPG Frühjahrstagung, Regensburg, Germany*, 2013.

[4] K. Ilin, D. Henrich, Y. Luck, L. Fuchs, J. M. Meckbach, and M. Siegel. Enhancement of critical current in mesoscopic superconducting strips by external magnetic field. In *DPG Frühjahrstagung, Regensburg, Germany*, 2013.

[5] M. Rudolph, J. Nagel, J. M. Meckbach, M Kemmler, K. Ilin, M. Siegel, D. Koelle, and R. Kleiner. Dc SQUIDs with asymmetric shunt resistors. In *DPG Frühjahrstagung, Regensburg, Germany*, 2013.

[6] T. A. Scherer, G. Aiello, G. Grossetti, A. Meier, S. Schreck, P. Spaeh, D. Strauss, A. Vaccaro, M. Siegel, J. M. Meckbach, and A. Scheuring. Reduction of surface losses of CVD diamond by passivation methods. In *International Conference on Infrared, Millimeter and THz Waves (IRMMW-THz 2012), Wollongong, Australia*, 2012.

[7] D. Henrich, P. Reichensperger, Y. Luck, L. Rehm, S. Dörner, J. M. Meckbach, M. Hofherr, K. Ilin, M. Siegel, A. Semenov, H.-W. Hübers, and D. Vodolazov. Effect of Bends on the Critical Current in Superconducting Nanowires. In *Applied Superconductivity Conference, Portland, USA*, 2012.

[8] J. M. Meckbach, M. Merker, S. J. Buehler, K. Ilin, B. Neumeier, U. Kienzle, E. Goldobin, R. Kleiner, D. Koelle, and M. Siegel. Sub-µm Josephson Junctions for Superconducting Quantum Devices. In *Applied Superconductivity Conference, Portland, USA*, 2012.

[9] S. J. Bühler, J. M. Meckbach, S. Wünsch, M. Merker, K. Ilin, B. Neumeier, E. Goldobin, R. Kleiner, D. Koelle, and M. Siegel. Creation of artificial 1D vortex-crystals using current injectors. In *Kryoelektronische Bauelemente 2012, Freudenstadt, Germany*, 2012.

[10] M. Merker, J. M. Meckbach, K. Ilin, and M. Siegel. Fabrication process for planarization and miniaturization of Josephson Junctions. In *Kryoelektronische Bauelemente 2012, Freudenstadt, Germany*, 2012.

[11] J. M. Meckbach, S. Bühler, M. Merker, K. Ilin, M. Siegel, K. Buckenmaier, T. Gaber, U. Kienzle, B. Neumaier, E. Goldobin, R. Kleiner, and D. Koelle. Versatile Multi-Layer Josephson Junction Process for Vortex Molecules. In *DPG Früjahrstagung, Berlin, Germany*, 2012.

[12] S. V. Shitov, A. A. Kuzmin, J. M. Meckbach, K. S. Ilin, S. Wuensch, A. V. Ustinov, and M. Siegel. Development of TES Bolometers with High-Frequency Readout Circuit. In *23rd International Symposium on Space Terahertz Technology, Tokyo, Japan*, 2012.

[13] J. M. Meckbach, K. Ilin, and M. Siegel. Technology of thin nitride films of Al and Nb for cryogenic quantum applications. In *Kryoelektronische Bauelemente 2011, Autrans (Grenoble), France*, 2011.

[14] T. Gaber, U. Kienzle, K. Buckenmaier, H. Sickinger, J. M. Meckbach, K. Ilin, M. Siegel, D. Koelle, R. Kleiner, and E. Goldobin. Thermal and quantum escape of fractional Josephson vortices. In *Kryoelektronische Bauelemente 2011, Autrans (Grenoble), France*, 2011.

[15] E. Goldobin, T. Gaber, K. Buckenmaier, U. Kienzle, H. Sickinger, D. Koelle, R. Kleiner, J. M. Meckbach, Ch. Kaiser, K. Ilin, and M. Siegel. Thermal and Quantum depinning of a fractional Josephson vortex. In *75. Jahrestagung der DPG und DPG Frühjahrstagung, Dresden, Germany*, 2011.

[16] J. M. Meckbach, Ch. Kaiser, W. Heni, S. Heunisch, K. Ilin, and M. Siegel. Nb/Al-AlO$_x$/Nb fabrication process for High-Quality Josephson Junctions. In *Kryoelektronische Bauelemente 2010, Zeuthen (Berlin), Germany*, 2010.

[17] K. Buckenmaier, T. K. Gaber, U. Kienzle, H. Sickinger, J. M. Meckbach, Ch. Kaiser, K. Ilin, M. Siegel, D. Koelle, and E. Goldobin. Activation energy of fractional Josephson vortices. In *Kryoelektronische Bauelemente 2010, Zeuthen (Berlin), Germany*, 2010.

[18] T. Schwarz, J. Nagel, J. M. Meckbach, K. Ilin, M. Siegel, R. Kleiner, and D. Koelle. Negative absolute resistance in annular Josephson junctions. In *Kryoelektronische Bauelemente 2010, Zeuthen (Berlin), Germany*, 2010.

[19] J. M. Meckbach, Ch. Kaiser, K. Ilin, M. Siegel, K. Buckenmaier, T. Gaber, U. Kienzle, H. Sickinger, E. Goldobin, R. Kleiner, and D. Koelle. Optimization of Nb/Al-AlO$_x$/Nb Technology for the Investigation of Fluxon Dynamics in Long Josephson Junctions. In *DPG Jahrestagung und DPG Frühjahrstagung des AKF, Regensburg, Germany*, 2010.

[20] Ch. Kaiser, J. M. Meckbach, S. Skacel, S. Wünsch, K. Ilin, and M. Siegel. Scalable Josephson Junction Technology for Quantum Devices. In *S-PULSE Workshop 2009 on Applied Superconductor Electronics in Quantum Metrology, Braunschweig, Germany*, 2009.

[21] J. M. Meckbach, Ch. Kaiser, K. Ilin, M. Siegel, K. Buckenmaier, U. Kienzle, T. Gaber, H. Sickinger, and E. Goldobin. High Quality Long Josephson Junctions and Realization of µm-sized Current Injectors for Investigations of Fractional Josephson Vortices. In *Kryoelektronische Bauelemente 2009, Oberhof, Germany*, 2009.

[22] H. Sickinger, K. Buckenmaier, U. Kienzle, T. Gaber, J. M. Meckbach, Ch. Kaiser, K. Ilin, M. Siegel, R. Kleiner, D. Koelle, and E. Goldobin. Spectroscopy of the eigenfrequencies of a fractional vortex molecule in a long annular Josephson junction. In *Kryoelektronische Bauelemente 2009, Oberhof, Germany*, 2009.

[23] Ch. Kaiser, J. M. Meckbach, K. Ilin, and M. Siegel. Scalable Josephson Junction Technology for Quantum Devices. In *European Conference on Applied Superconductivity, Dresden, Germany*, 2009.

Karlsruher Schriftenreihe zur Supraleitung
(ISSN 1869-1765)

Herausgeber: Prof. Dr.-Ing. M. Noe, Prof. Dr. rer. nat. M. Siegel

Die Bände sind unter www.ksp.kit.edu als PDF frei verfügbar oder als Druckausgabe bestellbar.

Band 001
Christian Schacherer
Theoretische und experimentelle Untersuchungen zur Entwicklung supraleitender resistiver Strombegrenzer. 2009
ISBN 978-3-86644-412-6

Band 002
Alexander Winkler
Transient behaviour of ITER poloidal field coils. 2011
ISBN 978-3-86644-595-6

Band 003
André Berger
Entwicklung supraleitender, strombegrenzender Transformatoren. 2011
ISBN 978-3-86644-637-3

Band 004
Christoph Kaiser
High quality Nb/Al-AlOx/Nb Josephson junctions. Technological development and macroscopic quantum experiments. 2011
ISBN 978-3-86644-651-9

Band 005
Gerd Hammer
Untersuchung der Eigenschaften von planaren Mikrowellen- resonatoren für Kinetic-Inductance Detektoren bei 4,2 K. 2011
ISBN 978-3-86644-715-8

Band 006
Olaf Mäder
Simulationen und Experimente zum Stabilitätsverhalten von HTSL-Bandleitern. 2012
ISBN 978-3-86644-868-1

Band 007
Christian Barth
High Temperature Superconductor Cable Concepts for Fusion Magnets. 2013
ISBN 978-3-7315-0065-0

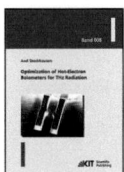

Band 008
Axel Stockhausen
Optimization of Hot-Electron Bolometers for THz Radiation. 2013
ISBN 978-3-7315-0066-7

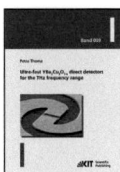

Band 009
Petra Thoma
Ultra-fast $YBa_2Cu_3O_{7-x}$ direct detectors for the THz frequency range. 2013
ISBN 978-3-7315-0070-4

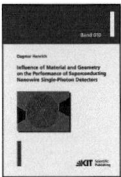

Band 010
Dagmar Henrich
Influence of Material and Geometry on the Performance of Superconducting Nanowire Single-Photon Detectors. 2013
ISBN 978-3-7315-0092-6

Band 011
Alexander Scheuring
Ultrabreitbandige Strahlungseinkopplung in THz-Detektoren. 2013
ISBN 978-3-7315-0102-2

Band 012
Markus Rösch
Development of lumped element kinetic inductance detectors for mm-wave astronomy at the IRAM 30 m telescope. 2013
ISBN 978-3-7315-0110-7

Band 013
Johannes Maximilian Meckbach
Superconducting Multilayer Technology for Josephson Devices. 2013
ISBN 978-3-7315-0122-0